United Universes
Quaternion Universe - Octonion Megaverse

Complex Quaternion Universe (QUeST)
Complex Octonion Megaverse (MOST)
New Hypercomplex Basis for Unified SuperStandard Theory
Real $\rightarrow$ Complex $\rightarrow$ Quaternion $\rightarrow$ Biquaternion $\rightarrow$ Octonion $\rightarrow$ Bioctonion
Darkness Quantum Number & SuperSelection Rule
Particle-Dimension Duality
Dimension-Functional-Particle Triality
Octoquarks
Megaverse Enhanced Dark Sectors
8-Dimensional Bioctonion Theory (MOST) Implies Unified SuperStandard Theory
MOST Differentiates the Normal and Dark Sectors of Universes and the Megaverse
MOST Octonion Spin-Darkness Connection
MOST Spinors Explains the Lack of Interactions with Dark Matter

Stephen Blaha Ph. D.
Blaha Research

Pingree-Hill Publishing
MMXX

Rev. 00/00/0
1   February 3, 2020

To Margaret
With Love

Some Other Books by Stephen Blaha

All the Megaverse! Starships Exploring the Endless Universes of the Cosmos using the Baryonic Force (Blaha Research, Auburn, NH, 2014)

SuperCivilizations: Civilizations as Superorganisms (McMann-Fisher Publishing, Auburn, NH, 2010)

All the Universe! Faster Than Light Tachyon Quark Starships & Particle Accelerators with the LHC as a Prototype Starship Drive Scientific Edition (Pingree-Hill Publishing, Auburn, NH, 2011).

Cosmos Creation: The Unified SuperStandard Model, Volume 2, SECOND EDITION (Pingree Hill Publishing, Auburn, NH, 2018).

Immortal Eye: God Theory: Second Edition (Pingree Hill Publishing, Auburn, NH, 2018).

Unification of God Theory and Unified SuperStandard Model THIRD EDITION (Pingree Hill Publishing, Auburn, NH, 2018).

The Exact QED Calculation of the Fine Structure Constant Implies ALL 4D Universes have the Same Physics/Life Prospects (Pingree Hill Publishing, Auburn, NH, 2019).

Unified SuperStandard Theory and the SuperUniverse Model: The Foundation of Science (Pingree Hill Publishing, Auburn, NH, 2018).

Quaternion Unified SuperStandard Theory (The QUeST) and Megaverse Octonion SuperStandard Theory (MOST) (Pingree Hill Publishing, Auburn, NH, 2020).

Available on Amazon.com, bn.com Amazon.co.uk and other international web sites as well as at better bookstores (through Ingram Distributors).

CONTENTS

FIGURES and TABLES

INTRODUCTION

Previous derivations of the Unified SuperStandard Theory were based on Complex General Relativity and Quantum Field Theory. This book presents a deeper derivation that unites space-time coordinates and internal symmetry coordinates. Our motivation is the close analogy between Lorentz subgroups and Standard model groups suggesting a possible common origin. We seek a corresponding structure in the form of a larger space-time.

The basis for our approach is the trend of types of coordinates:

Real → complex → quaternion → Complex quaternion → octonion → Complex octonion

The Standard Model is based on real-valued coordinates. The Unified SuperStandard Theory, and its precursors by the author, were based on complex coordinates (specifically Complex Lorentz group transformations). Recently the author has developed biquaternion and bioctonion SuperStandard Models in Blaha (2020) with features that are very similar to the Unified SuperStandard Theory previously developed by the author. This book continues that development.

It creates a combined SuperUnified theory that unites the biquaternion theory for our universe (QUeST) with the Megaverse bioctonion theory (MOST) described in Blaha (2020).

The new theory meshes well with the universe and Megaverse forms described previously. They all are consistent with the Unified SuperStandard Theory. They contain it and add additional features. Thus they are supersets of the Unified SuperStandard Theory. They have many new noteworthy features.

Quaternions and octonions are hypercomplex numbers with special properties that make them similar to complex numbers. Quaternions and octonions are both normed division algebras over the reals (*hypercomplex* number systems) with salutary properties for quantitative studies in quantum field theory and perturbation theory. Some of their new features are listed on the cover page.

We will begin with a new, more primitive, set of axioms that lead to the theories. Then we proceed to develop the biquaternion universe theory followed by the bioctonion theory. Subsequently we explore features and the unification of the respective theories with our universe residing within the Megaverse. We suggest the Megaverse has a larger set of (Dark) fermions, vector bosons and interactions.

We find MOST has three sets of Dark sectors that have a Darkness superselection rule. This rule precludes interactions amongst the normal and three Dark sectors. We further find a fermion – dimension duality that leads to a particle – functional – dimension triality.

In MOST the 16-spinors of 8-dimensional space-time "divide" into four 4-spinors – one 4-spinor for each of the four sectors. This feature carries over to QUeST and the Unified SuperStandard Theory to give a spin-based justification for the lack of interactions between normal and Dark matter. The Darkness quantum number superselection rule systemizes the non-interaction.

An exciting new feature of MOST is a new type of 8-dimensional quark that we call octoquarks. Octoquarks have a real-valued momentum and spin but they also have seven hypercomplex momenta.

1. New Axioms

It seems that the best set of axioms is the set that is most general in its statements, and most comprehensive in its consequences. In this chapter we present a new, more primitive, set of axioms for the SuperUnified Quaternion Universe/Megaverse Octonion SuperStandard Theory. We then outline its consequences. It will become clear that the set of axioms leads to the derivation of the theory. Remarkably, it will also become clear that a unifying framework, based on one general principle, does not seem to be possible.

This point is particularly evident if one considers the determination of coupling constants (such as those of the Standard Model). We showed that the electromagnetic fine structure constant is exactly determined by vacuum polarization. We also showed the ElectroWeak and the Strong interaction coupling constants are also determined to good accuracy by vacuum polarization calculations. Thus unification of these interactions does not appear to be relevant physically.

As a result a set of axioms appears to be the only reasonable basis for the derivation of a complete theory of elementary particles and cosmology.

1.1 Axioms of the SuperUnified Quaternion Universe/Megaverse Octonion SuperStandard Theory

In this section we state a new set of axioms that are more general than those of the Unified SuperStandard Theory. They imply those of the Unified SuperStandard Theory. They assume that a hypercomplex form of General Relativity holds.

AXIOMS

1. Biquaternion space is the basic space-time of our universe. Bioctonion space is the basic space-time of the Megaverse. These spaces factorize into a coordinate space-time and an internal symmetry space-time.

2. Physical processes can execute in parallel.

3. Matter and energy are particulate.

4. Space--times are locally Lorentzian.

5. All calculations are finite.

6. Particle theory can be defined in any curved space-time.

7. Each particle has a wave function determined by a functional inner product defining the particle state. The functionals form a set without a distance measure.

1.2 General Implications of the Axioms

In this section we describe some of the implications of each of the axioms.

1. Biquaternion space is the basic space-time of our universe. Bioctonion space is the basic space-time of the Megaverse. These spaces factorize into a coordinate space-time and an internal symmetry space-time.

The factorization into a space-time and an internal symmetry space must be a form of spontaneous symmetry breaking of yet unknown origin. It appears to be related to a breakdown of the vacuum.

2. Physical processes can execute in parallel.

Physical processes are known to be able to execute in parallel at any distance of separation. As Fant has shown parallel execution requires a minimal number of dimensions: 4. Consequently the dimension of space-time must be 4 or greater. The biquaternion space-time of QUeST is 4-dimensional allowing parallel process execution.

The bioctonion space-time of MOST is 8-dimensional and also allows parallel process execution. The choice of eight dimensions is natural since it allows 4-dimensional universes within it. It also has a form that allows a clean formulation. Lastly, as will be seen later, it conforms to the pattern of interplay between Lorentz symmetry and internal symmetry found in the Unified SuperStandard Theory. See Appendix 1-A for an extended discussion of parallelism and dimension.

This axiom leads to a view of the origin of the dimensions.

3. Matter and energy are particulate.

The most direct method of specifying a theory of matter and energy is through the Use of Quantum Field Theory. Thus Quantum Field Theory is implied.

4. Complex Space-times are locally Lorentzian.

A locally complex Lorentzian space-time leads to Complex General Relativity. In flat space-time Complex General Relativity becomes Complex Lorentz group. (In point of fact the Complex Poincare group follows.)

5. All calculations are finite.

Given the need for Quantum Field Theory it becomes necessary to find a formulation that yields finite values for calculations in perturbation theory. The only approach that eliminates high energy divergences, and yet preserves the results found in perturbation theory calculations that agree with (primarily QED) experiments, is Two-Tier Quantum Field Theory. This is discussed in detail in earlier books starting in 2002. Thus only our Two-Tier formalism satisfies this axiom.

6. Particle theory can be defined in any curved space-time.

In the 1970s we developed a formalism that allows the definition of particle states in any space-time in such a way that its physical content is preserved when transformed to any coordinate system.[1] This PseudoQuantum Quantum Field Theory satisfies this axiom.

7. Each particle has a wave function determined by a functional inner product defining the particle state. The functionals form a set without a distance measure.

This axiom is satisfied by our formulation of quantum functionals in Blaha (2019f) and earlier books. Our formulation eliminates the superficial violation of the Theory of Relativity by "spooky" quantum entangled processes with parts separated by a physically "large" distance.

The seven axioms imply the Unified SuperStandard Theory and its deeper biquaternion and bioctonion hypercomplex formulations.

---

[1] S. Blaha, Il Nuovo Cimento **49A**, 35 (1979).

1.3 Hypercomplex General Relativity

In the case of hypercomplex coordinate systems such as a quaternion-based system or an octonion-based system one can define a Hypercomplex General Relativity along the lines of General Relativity based on real-valued coordinates. In doing so one must be mindful of issues of commutativity and associativity for hypercomplex-valued coordinates.

1.4 Complex Special Relativity

As we shall see we find hypercomplex space factorizes into complex-valued space-time in 4 or 8 dimensions for biquaternions and bioctonions respectively. Then one can create a Complex General Relativity in four or eight dimensions that becomes a Complex Special Relativity in flat space-time.

Appendix 1-A. Space-Time Dimensions

Generally the dimension of space-time is taken as given and inquiries are not made as to the origin and purpose of the number of dimensions. In this appendix we show a fundamental justification exists for the number of space-time dimensions. The justification is simply the physical requirement that (asynchronous) physical processes can execute in parallel.

1-A.1 Determination of the Number of Space-Time Dimensions

The *a priori* determination of the dimensions of a space-time is guesswork in the absence of a fundamental principle(s).

This appendix derives the dimensions of our 4-dimensional space-time using one principle that yields the same space-time dimension of four. The principle (axiom) specifies the requirement that space-time must allow physical processes to run in parallel following the principles of Asynchronous Logic.[2] This requirement is physically necessary. Spatially separated quantum entangled processes must have parts that proceed in parallel at a distance. This principle is also required in order to have spatially separated classical processes to proceed in parallel.

1-A.2 Synchronization of Non-Local Physical Processes

In earlier books we discussed the central role of Logic and the need for synchronization of non-local physical processes. The need for synchronized non-local physical processes requires the introduction of a new principle: the Principle of Asynchronicity.[3] When processes take place in parallel whether it is Quantum

---

[2] This approach is developed in (2012a) and (2015a).
[3] Much of this chapter is covered in Blaha (2011c) and so is printed in smaller type. Some might argue that it should be called the principle of synchronicity since the goal is synchronization of the parts of an evolving process. We chose to follow the terminology in the field of Asynchronous Logic as exemplified by Fant (2005) – a classic in that field.

Mechanical entangled processes at small/large distances, or in high order Feynman diagrams (or their old fashioned time ordered perturbation theory predecessor) the synchronicity of a process is a physical requirement. It is implicitly resolved by physical laws which prevent asynchronicities (situations when parallel processes get "out of sync" resulting in the failure of an entire physical process to complete properly.) The Principle of Asynchronicity is described in the following pages. Asynchronicity can be briefly described as:

> In computation asynchronicity issues can arise. For example parallel computations or computer processes on a chip or set of chips have to be carefully managed for a parallel computer process to complete properly. In the case of computer chip design (VLSI chips and so on) techniques have been developed for the design of chips based on multi-valued logic. One conceptual approach uses 4-valued logic to define clock-less computer logic circuits. The 4-valued logic developed by Fant (2005) has the four logic values TRUE, FALSE, NULL, and INTERMEDIATE. It is an extension of Boolean Logic that can accommodate time asynchronicities in asynchronous computer circuits. It enables circuits to avoid the use of system clocks to implement synchronization.[4] Thus the synchronization is explicit in 4-valued logic and non-logical constructs are not needed.[5] Concurrent transitions are coordinated solely by logical relationships with no need for any time constraints or relationships.

Realizing that the Standard Model, and physical theories that are ultimately derived from it, such as Quantum Mechanics, potentially contain asynchronicities, we suggest that a Principle of Asynchronicity is necessarily embodied in the fundamental theory of Physics.

1-A.3 Four-Valued Asynchronous Logic

The basic defining features of asynchronous circuits and Asynchronous Logic are:

---

[4] Remarkably Bjorken (1965) pp. 220-226 presents an analogy between Feynman diagrams and electrical circuits where momenta map to currents, coordinates to voltages, Feynman parameters to resistance, and free particle equations of motion to Ohm's Law plus the equivalent of Kirchhoff's Laws. Thus Feynman diagrams and computer circuits are analogous.
[5] A two-valued asynchronous logic is also possible – just as the Dirac equation can be expressed as two 2-dimensional equations. See Fant (2005) and Bjorken (1965).

1. An *asynchronous circuit* is a circuit in which the component parts are autonomous and can act in parallel at various rates of time evolution. They are not controlled by a clock mechanism but proceed, or wait for signals indicating that they can proceed.

2. *Asynchronous logic* is the logic used in the design of asynchronous circuits. The logic embodies the asynchronicity, and so the circuits built using the logic do not use a clock to control the execution speed of the various parts of an asynchronous circuit. Consequently logic elements do not necessarily have a distinct true or false state at any given point in time. 2-valued Boolean Logic is not sufficient and so asynchronous logic is multi-valued. The logic contains states that allow for "stop and go" states within an executing asynchronous circuit.

In Fant's asynchronous 4-valued logic the four possible truth values of a state are:

> True – status is true and all data is current
> False – status is false and all data is current
> Intermediate – status is indefinite with some data current
> NULL – status is indefinite with no data present – results in a suspension of processing of the circuit part in a NULL state until current data becomes present

"Data" is the information flowing through all or part of a circuit. Using these truth values the evolution in time of the parts of an asynchronous circuit are effectively synchronized by the logic without the use of a clock mechanism. (A clock mechanism effectively is a subsidiary time constraint or set of time constraints.) See Fant (2005) for further details.

An implicit aspect of asynchronous logic is the coordination of spatially separated parts of a circuit. Since spatial separations in a circuit can be mapped to time delays using the speed of data propagation between parts, spatial asynchronicites are subsumed under time asynchronicities. This is particularly true for computer chips which are kept small to minimize delays.

1-A.4 Principle of Asynchronicity

An obvious feature of elementary particle phenomena is the coordination of the parts of a physical process in time and space. Complex Feynman diagrams embody the coordination of the parts of interacting particles. Quantum entanglement phenomena embody the coordination of the parts of a physical phenomenon separated by large distances and perhaps times. Examples of these types, which could be multiplied indefinitely, lead to a Principle of Asynchronicity.

Principle (Axiom): Nature requires an asynchronicity principle. Aasynchronicity is coordinated by 4-valued physico-logical structures for matter.

Elaboration: Elementary particle physical phenomena must support extended coordinated physical phenomena in space and time. The fundamental laws of particle physics must be such as to permit coordinated physical phenomena with coordination between the parts of a physical phenomenon at small/large distances and small/large time intervals. The coordination must be embodied within physical laws. We make it an axiom of our theory.

Asynchronicities are common in the many sub-circuits of a computer chip. Asynchronicities are also common in the many interaction sub-regions of a set of interacting particles. Page 7 of Fant (2005) has a diagram of a circuit with a set of sub-circuits with five time slices of the interacting sub-circuits showing five states of the "'data' wave front" at five points in time. This diagram is similar to the time-sliced diagram of an interacting system of particles in "old fashioned" time-ordered perturbation theory. Page 29 of Blaha (2005b) displays a similar diagram (Fig. 5.1.4) in a description of a Standard Model Quantum Language Grammar – a language representation of particle physics. Blaha's diagram[6] is remarkably similar to Fant's diagram in overall features as one might expect since both address time asynchronicity.

The asynchronicity that appears in perturbation theory diagrams is intimately related to the appearance of antiparticles in diagrams. As noted earlier antiparticles are

---

[6] Created without knowledge of Fant's work.

interpretable as negative energy particles traveling backwards in time. The time orderings, which are implicit in the Feynman diagram approach and explicit in old fashioned perturbation theory, evidence time asynchronicity and the effects of the dynamics. They coordinate asynchronicities such that correct results follow from perturbative calculations.

1-A.5 Dimension of the Megaverse

Having found the dimension of our universe is 4, we now turn to the dimension of the enclosing Megaverse. The dimension must be greater than four. But its value cannot be specified by a fundamental principle except perhaps minimality. Later we will see the choice of a dimension of eight yields a Megaverse set of internal symmetries that are a simple superset of the dimensions of our 4-dimensional QUeST. It does add additional Dark sectors in a way that is compatible with the Unified SuperStandard Theory.

Other choices such as a 6-dimensional alternative are either more complex or have too many internal symmetries and fundamental particles.

On this basis we choose an 8-dimensional Megaverse. It gives a Megaverse that can have an infinite number of universes and yet is not "overly" wasteful of particles and interactions.

Appendix 1-B. The Previous Basis of the Unified SuperStandard Theory

It seems worthwhile to review the previous basis of the Unified SuperStandard Theory because the new deeper basis builds on it. This appendix is an extract from Blaha (2019g) and (2020).

1-B.1 A Foundation in Complex General Relativity

We now turn to follow the path that ultimately leads to the SuperStandard Theory. Unlike other attempts at a fundamental theory we do not posit groups for interactions but rather provide strong arguments (derivations) for the choice of the Standard Model group and the SuperStandard Theory group.

We assume that the fundamental theory of space-time coordinates is Complex General Relativity. It is General Relativity extended to complex-valued coordinates with a complex-valued energy-momentum tensor. Complex General Relativity can be factored into a U(4) group that rotates complex coordinates and a residual Complex General Relativity factor. We call the U(4) group the Coordinates Species group for reasons given later.

From Complex General Relativity we proceed to consider the flat space-time case with Complex Special Relativity. Complex Special Relativity is described by the Complex Lorentz Group which has the subgroups SU(2), U(1), SU(3), and additional SU(2) and U(1) subgroups.

After showing a map from coordinate subgroup symmetries to elementary particle group symmetries we find the Coordinates Species group maps to a U(4) Particle Species group.

1-B.2 A Foundation in Quantum Field Theory

Particles, and particle symmetries, to which we have been alluding above, emerge from Quantum Field Theory—the other Foundation of Physics. Why do we view Quantum Field Theory as a fundamental foundation? Of all the forms a dynamics theory might take, Quantum Field Theory is the simplest form that supports the creation and annihilation of matter—a fundamental attribute of matter as we know it. Blob creation and annihilation would take us into complexity as would realistic string creation and annihilation.

Quantum Field Theory offers a simplicity that is easily seen in the representation of creation and annihilation using Feynman diagrams.

Thus we opt for Quantum Field Theory and find it, and Complex General Relativity, sufficient to describe all known features of elementary particles and their combinations into more complex forms of matter and energy. As we saw in Blaha (2019g) and (2018e) and our earlier books the purported problems of Quantum Field Theory are easily curable.

A direct benefit of Quantum Field Theory is the appearance of particle number operators which lead to the Generation Group and the Layer Groups

1-B.3 Axioms of the Unified SuperStandard Theory

Complex General Relativity and Quantum Field Theory lead to the Unified SuperStandard Theory. The fundamental axioms that specify the basis of the derivation of the Model are listed in Fig. 1-B.1.[7] Their detailed implications are:

1. Each space-time symmetry subgroup of the Complex Lorentz Group has a corresponding particle interaction symmetry group. The particle symmetry groups combine in a direct product.

    The specific subgroups of the Complex Lorentz groups with corresponding particle interaction symmetry groups is only well-defined if we further require that they correspond to the distinction between space and time. Thus the SU(3)

---

[7] Chapters 6 and 13 provide a revised deeper set of axioms for our universe and the Megaverse respectively.

subgroup emerges from the 3 × 3 space part of Complex Lorentz Group elements. The pair of SU(2) ⊗ U(1) subgroups emerge from the "boost" parts of Complex Lorentz Group elements. Consequently the corresponding enlarged Standard Model particle interaction symmetry is

$$SU(2) \otimes U(1) \otimes SU(3) \otimes SU(2) \otimes U(1)$$

It is the minimal group where the above direct product is a subgroup of SU(7).

2. Quantum Field Theory supports fundamental particles that form a countable set. Each particle number operator is a generator of a particle interaction group.

All matter and energy is composed of discrete particles.

3. All quantum field theory calculations are finite.

Perturbation theory calculations have divergences in conventional quantum field theory that require renormalization. Axiom 3 below leads to the author's Two Tier formulation[8] where there are no divergences—eliminating the need for renormalization to eliminate divergences.

4. The Quantum Field Theory of particles can be defined in any curved space-time.

In certain curved space-times conventional second quantization leads to ambiguities in the definition of particle states. Axiom 4 below leads to the author's generalized second quantization procedure,[9] called PseudoQuantization, which eliminates these ambiguities. It also supports a canonical lagrangian formulation of higher derivative field theories.[10]

---

[8] See Blaha (2005a).
[9] S. Blaha, Phys. Rev. D**17**, 994 (1978) and references therein to earlier papers by the author such as Phys. Rev. D**10**, 4268 (1974) and Il Nuovo Cimento **49A**, 35 (1979) and **49A**, 113 (1979)
[10] S. Blaha, Phys. Rev. D**10**, 4268 (1974) and D**11**, 2921 (1975) and references therein.

5. Each particle wave function has a quantum functional[11] defining the particle state in a space without a distance measure.

There is a set of functionals (called monads or cores) with an element for each particle in the universe.[12] The quantum entanglement of particles at a distance can be instantaneous because the functionals (which embody the state of each particle) exist in a space without a distance measure. In a sense this feature embodies the unity of creation.

The five axioms imply the detailed list of axioms in Blaha (2019g) and (2018e).The axioms are revised for QUeST (chapter 6) and also for MOST (chapter 13).

---

[11] See Blaha (2019e) and (2018f) for a detailed discussion. Our approach eliminates the issues of "spookiness" and instantaneous action at a distance that clouds quantum entanglement.
[12] See Blaha (2019e) and (2018f).

2. Summary of Features of the Unified SuperStandard Theory

This chapter summarizes *some* features of the Unified SuperStandard Theory (the Theory) for comparison with the more fundamental biquaternion theory QUeST and the bioctonion theory MOST. The Theory was based in part on the complex (and real) coordinates of the Complex Lorentz group and more deeply on Complex General Relativity and Quantum Field Theory.

The detailed discussion of the Theory is presented in Blaha (2018e) as extended and amended by Blaha (2019f). Parts of these books appeared in refereed papers in the 1970s and in books .starting with Blaha (2002). See the list of book References near the end of this book.

2.1 Internal Symmetry Groups

The internal symmetry groups of the Theory include the $SU(2) \otimes U(1)$ ElectroWeak theory, $SU(3)$, and (badly broken) groups based on particle number operators. The complete internal symmetry group is

$$[U(1) \otimes SU(2) \otimes SU(3) \otimes U(1) \otimes SU(2) \otimes SU(3) \otimes U(4)^4]^4 \otimes U(4) \qquad (2.1)$$

where the last $U(4)$ factor is the Species group. The two factors of $U(1) \otimes SU(2) \otimes SU(3)$ are for normal matter and Dark matter. The $U(4)^4$ factor is the product of Generation groups and Layer groups for normal and Dark matter totaling four factors. The exponent 4 outside the brackets reflects the four layers of particles and groups of the Theory.

2.2 Fermion Spectrum

The fermion spectrum consists of four layers, each containing four generations of normal and Dark fermions. It is derived in Blaha (2019e) and earlier books by the author. The spectrum appears in Fig. 2.1.

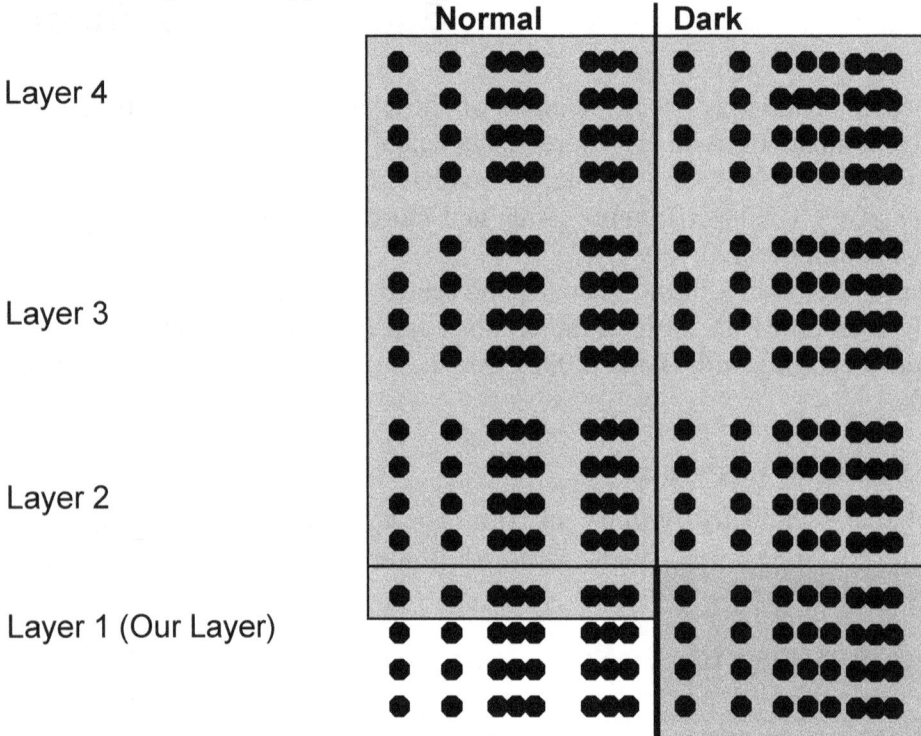

Figure 2.1 The fundamental fermion spectrum consists of four layers of four generations of fermions. Current unknown (meaning not yet found experimentally) matter parts of the periodic table are grayed. Light parts are the known fermions with an additional, as yet not found, 4th generation shown. The Theory further distinguishes normal and Dark above. The Normal and Dark matter sections are of similar form. There are 256 fundamental fermions.

2.3 Vector Boson Spectrum

The vector bosons (interactions) of the theory are shown in Figs. 2.2 and 2.4.

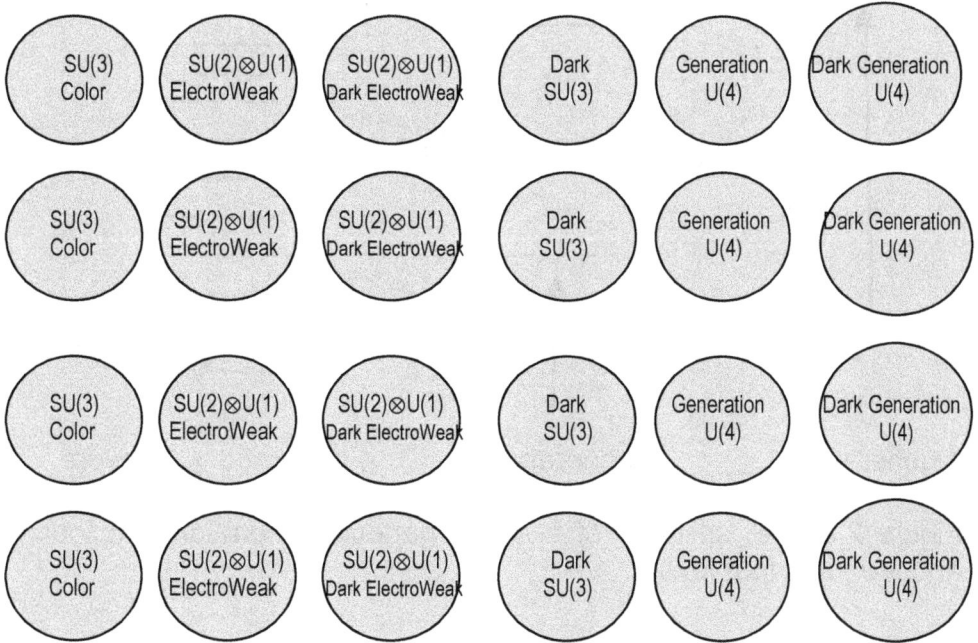

Figure 2.2. The set of four layers of internal symmetry groups corresponding to four generations in four layers of spin ½ fermions and the four layers of vector bosons. In addition there are the Normal and Dark Layer groups, the Species group and the Interaction Rotations group Θ that are *not* displayed.

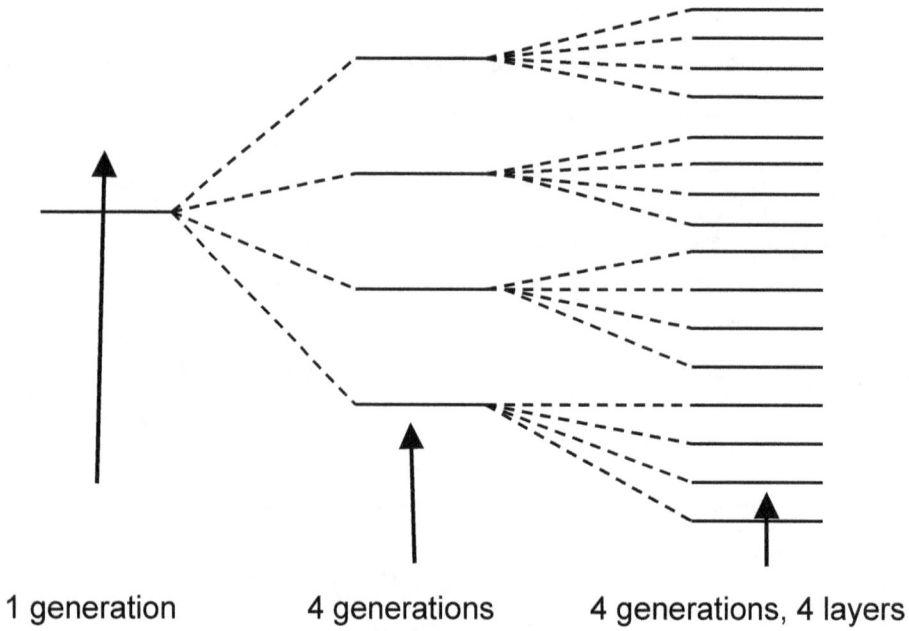

1 generation   4 generations   4 generations, 4 layers

Figure 2.3 The "splitting" of a single generation fermion into four generations and then into four layers.

The Vector Boson "Periodic" Table

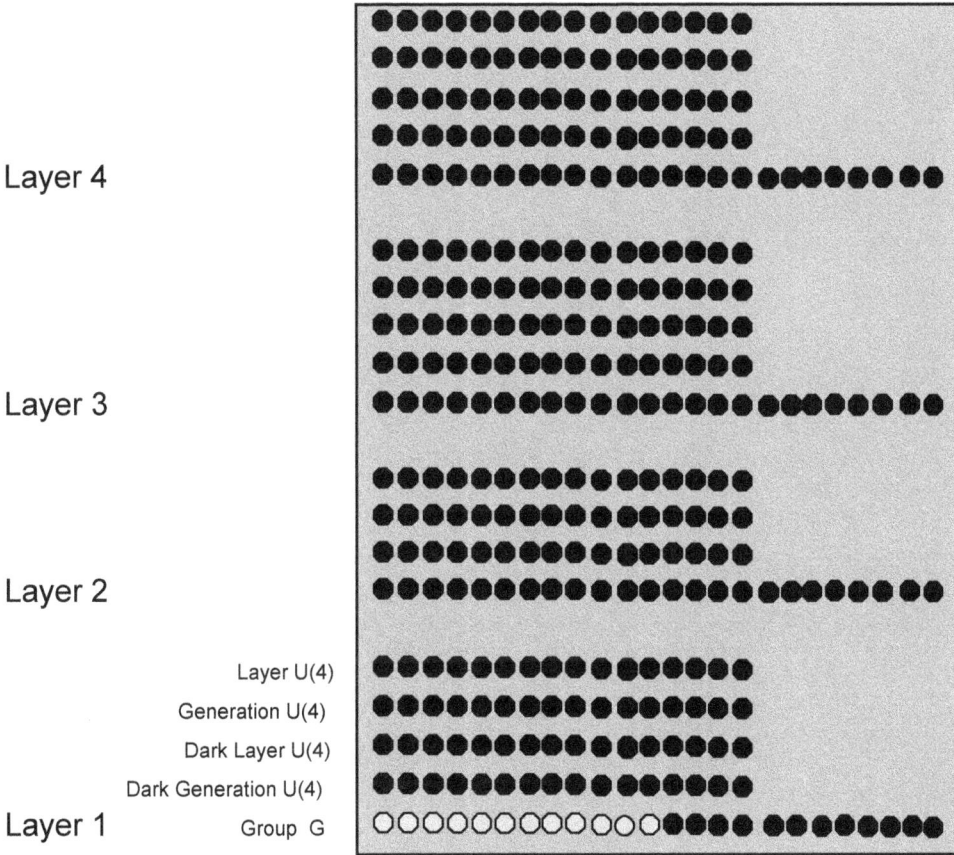

Figure 2.4. The vector bosons. Each circle represents a group generator. The known vector bosons are in the lowest row with a white interior. Yet to be found vector bosons are solid black. The Layer groups are distributed by layer symbolically although they each straddle all four layers. G is $SU(3) \otimes SU(2) \otimes U(1) \otimes SU_D(2) \otimes U_D(1) \otimes SU_D(3)$. The list of groups for the higher three levels is the same as those of the first layer. There are 352 vector bosons.

3. Quaternion Unified SuperStandard Theory (QUeST) Formulation Compared to Unified SuperStandard Theory

This chapter[13] lists some of the highlights of the Unified SuperStandard Theory (presented in Blaha (2018e) and (2019g)) and briefly identifies the differences from our new Quaternion Unified SuperStandard Theory (QUeST) formulation presented in this book for the first time. QUeST gives a deeper foundation. It has the benefit of corresponding to the known features of the Standard Model and most of the features of the Unified SuperStandard Theory. *It solidifies the symmetry structure of the theory into a concrete pattern and thus avoids the use of conjectural ansätze to determine elementary particle symmetries.*

1. The number of spatial dimensions was determined to be the number of generators in the primary set of interactions of the space. In the case of an *empty* universe the primary set of interactions is the U(2) qubit transformations group. The number of U(2) generators is four and thus the dimension of space is 4 complex dimensions. Also and more importantly, considerations of Asynchronous Logic, and the requirement that physical processes must be able to proceed in parallel, require the number of spatial dimensions to be four. The book justifies four complex space-time dimensions with a Lorentz metric yielding Complex Lorentz group symmetry.

Change: The dimensionality is set by the quaternion foundation..

2. Boosts of the Complex Lorentz group transform a Dirac-like equation with a Landauer mass into four different forms (called species). Each form maps to a type of

---

[13] This chapter is extracted from Blaha (2020) for the reader's convenience.

fermion: neutral leptons (neutrinos), charged leptons, up-type quarks, and down-type quarks. Neutral leptons and down-type quarks are tachyons. Some evidence exists for tachyonic neutrinos. Complex Lorentz boosts lead to the Complex Lorentz group factorization: $SU(2) \otimes U(1) \otimes SU(3) \otimes SU(2) \otimes U(1)$. We map $SU(2) \otimes U(1) \otimes SU(3)$ to fermion particle functional space to obtain the internal symmetry group for ElectroWeak and Strong Interactions: $SU(2) \otimes U(1) \otimes SU(3)$. The remaining factors $SU(2) \otimes U(1)$ we map to the internal symmetry group for Dark Matter, which we take to be the Dark ElectroWeak Interaction (unconnected to normal matter interactions).

> Change: We find an additional SU(3) Dark Strong Interaction group giving a $SU(2) \otimes U(1) \otimes SU(3) \otimes SU(2) \otimes U(1) \otimes SU(3)$ symmetry.

3. Parity Violation as seen in the Weak Interactions follows directly from the forms of the four types of fermions predicted by the Complex Lorentz Group.

> Change: None

4. The existence of four conserved (and partially conserved) quantum numbers such as baryon number and lepton number indicates that there is a U(4) group whose **4** representation causes each species to have four generations—three of the generations are known. We suggest that a fourth generation of much higher mass fermions exist.

> Change: Normal matter has a U(4) Generation group and Dark matter has a separate U(4) Dark Generation group.

5. In each generation there are four partially conserved quantum numbers. Thus we find that there is another U(4) group (called a Layer group) for each generation yielding the combined Layer groups $[U(4)]^4$. The **4** representation of each U(4) results in a fermion spectrum of four layers of four generations or 192 fermions in all. We see only one layer at present. The additional three layers of fermions remain to be found at much higher masses. The symmetry group of the Unified SuperStandard Theory is

$$[SU(2)\otimes U(1)\otimes SU(3)\otimes SU(2)\otimes U(1)\otimes U(4)\otimes U(4)]^4\otimes U(4)$$

where the last factor is for the broken Species group, which follows from Complex General Relativity.

Change: Changed symmetry. Additional Dark Strong SU(3) group and separate Generation and Layer groups for the Dark sector to avoid generating normal particle and Dark particle interactions. The result:

$$[U(1)\otimes SU(2)\otimes SU(3)\otimes U(1)\otimes SU(2)\otimes SU(3)\otimes U(4)^4]^4\otimes U(4)$$

6. Assuming all particles are massless at the Big Bang, and all particle types have an equal proportion of the total mass-energy then, we find that the 192 fermions and 192 vector bosons yield a Dark Matter percentage of 83.33% (experimentally the estimates are 84.5% and 81.5%). The proportion of Dark Mass-Energy is found to be 91% of the universe's mass-energy. Experimentally the proportion has been estimated to be 95%. These results agree well with experiment. See chapter 14 of Blaha (2019g) and (2018e) for details.

Change: Now 128 Dark fermions and 128 "normal" fermions (of which 24 are known) totaling 256 fermions. Now there are 340 Dark vector bosons and 12 known vector bosons total 352 vector bosons. As a result we find the percent of Dark matter is 90.6%, and the percent of Dark Mass-Energy is 96.6%., which are similar to the known data values. See Figs. 2.1 and 2.4.

7. The instantaneous quantum effects between space-like separated parts of a quantum state ('spookiness') are taken to be a feature of fundamental importance. The only sensible way to implement this feature in quantum theory is to assume that the wave function of every particle is the inner product of a particle functional and a wave (Fourier) coordinate expansion. Particle functionals exist in a space with no distance measure. The space of coordinate expansions also has no distance measure. Other functionals in a state (and their implicit coordinate Fourier expansions) change

instantaneously when one of the functionals comprising a state changes since coordinate space distance is irrelevant.

   Change: None

   8.   Fermion particle functionals are called *Qubes*. They exist 'within' every fermion. They have a mass that we take to be the Landauer mass—the minimal energy of a qubit. Boson particle functionals are called *Qubas*. They are assumed to be massless in the absence of all interactions to preserve free vector boson and spin 2 boson gauge symmetry. Free Higgs particles are assumed to be massless for consistency.

   Change: None

   9. To have a completely finite theory with no infinities (including no fermion triangle infinities) we introduced Two-Tier Coordinates that replaced normal point like coordinates with a type of 'fuzzy' coordinates.

$$X^\mu = x^\mu + iY^\mu(x)/M_c^2.$$

   Change: None

   10. Since the Unified SuperStandard Theory lagrangian would require higher order derivatives to account for quark confinement (linear potential terms) and for MoND-like deviations from conventional gravity, and since such terms would be outside a canonical lagrangian formulation, we introduced two fields for each particle (fermions and bosons) in a formulation we call PseudoQuantum Theory. PseudoQuantum theory enables a canonical lagrangian formulation. It has other advantages such as a clean separation of vacuum expectation values from quantum fields for Higgs particles. It also supports second quantization in arbitrary coordinate systems while maintaining the same particle interpretation of states in all coordinate systems.

Change: None

11. The book also describes Higgs symmetry breaking and the use of the Faddeev-Popov Mechanism in detail for the theory.

Change: None

12. Since a Complex Special Relativity requires a Complex General Relativity we considered Complex General Relativity and showed that it could be 'factored' into General Relativity and a new U(4) group that we called the Species group. Since Complex General Relativity must support interactions with all types of matter we specified a Species group interaction with all matter. Further, we assumed that the Species vector bosons acquired masses through the Higgs Mechanism. The Higgs Mechanism caused Species group contributions to each fermion mass. Such a mass term would require each fermion particle mass to be both inertial *and* gravitational *solving the mystery of the equality of inertial and gravitational mass.*

Change: None

13. We showed that the implicitly higher derivative Riemann-Christoffel curvature tensor for all interactions leads to new interactions beyond The Standard Model. In addition to yielding quark confinement and MoND-like modifications of gravity, it may help understand the missing nucleon spin issue, discrepancies in proton radius measurements, vector meson dominance (VDM), and so on.

Change: Addition of Dark SU(3) terms and Dark Generation and Layer groups to the Riemann-Christoffel tensor for each layer.

14. We defined an Interaction Rotations group that caused rotations among all the vector boson interactions of The Unified SuperStandard Theory. We found that rotations that respected Superselection rules such as the Charge Superselection rule

could have physical significance. One example is ElectroWeak Theory which is an application of Interaction Rotation transformations.

> Change: The Interaction Rotations group is now not compelling since it is not in QUeST.

15. Since the number of fundamental fermions (192) and fundamental vector interaction bosons (192) is equal we considered Supersymmetric features of the Unified SuperStandard Theory.

> Change: The number of fermions and vector bosons is changed. (Item 6.) SuperSymmetry not indicated.

16. The discovery of two new particles that do not appear to be within the framework of The Standard Model, as it is currently known, raises the possibility that they may be within the expanded fermion spectrum in The Unified SuperStandard Theory. Towards that end we present a *preliminary* assignment of the locations of the new fermions within the spectrum of the Unified SuperStandard Theory.

> Change: The additional Dark Strong SU(3) group implies Dark quarks are triplets.

17. We showed that the coupling constants of the Standard Model including the Fine Structure Constant of QED are determined by eigenvalue functions. As a result they differ and in a way that appears contrary to the hypotheses of Grand Unified Theories (GUTs) even taking account of running coupling constant considerations.

> Change: None

Appendix 3-A. Gauge Fields Based on Particle Numbers

In this Appendix[14] we show the origin of the Generation and Layer groups in particle number operators. Particle interactions followed directly in the Unified SuperStandard Theory by analogy with Complex General Relativity subgroups yielding

$$SU(2) \otimes U(1) \otimes SU(3) \otimes SU(2) \otimes U(1) \otimes SU(3) \qquad (3\text{-A.1})$$

where the latter three factors are the Dark interactions.

They have a SU(10) covering group that contains this direct product of groups. The groups in eq. 3-A.1 are particle interaction groups in the Unified SuperStandard Theory.

Unlike other attempts to develop a formulation of the Standard Model (or generalizations) the Unified SuperStandard Theory was originally directly based on a theory foundation consisting of Complex General Relativity and Quantum Field Theory. Later we will show a deeper basis in Quaternions and Octonions.

To those who might prefer to base a theory on real General Relativity we note that proofs in Quantum Field Theory *require* the Complex Lorentz Group.[15] Thus the Complex Lorentz group is unavoidable for a properly (and rigorously) formulated Quantum Field Theory. Since the formulation of the Complex Lorentz Group in flat space-time can only be as the limit of Complex General Relativity, the choice of a foundation of Complex General Relativity is required.

Since particles are countable, and thus have discrete particle numbers, Quantum Field Theory brings particle numbers, and particle number laws such as particle conservation laws, into consideration.

---

[14] This appendix is an extract from Blaha (2020 for the readers convenience.
[15] Streater (2000).

Blaha (2019e) and earlier books showed that Complex Lorentz boosts generate four types of fermion particles that we call *particle species*. We map these four species to charged leptons (such as electrons), neutral leptons (such as neutrinos), up-type quarks (such as the u quark), and down-type quarks (such as the d quark).

3-A.1 Basis of the Generation Group

We define two particle number operators for normal up-quark particles and down-quark particles, B_{uq} and B_{dq}. Similarly we define two particle number operators for normal species "e" (electron) particles and species "v" particles, B_e and B_v. Similarly we define Dark matter equivalents:[16] B_{De}, B_{Dv}, B_{Duq}, and B_{Ddq}.

In the absence of interactions these fermion particle number operators are conserved. Each set are "diagonal" operators within a U(4) group. Thus we have a normal U(4) Generation Group and a Dark U(4) Generation group.

On this basis we find there are four generations of each species in the normal and in the Dark matter sectors. One generation of normal fermions with large masses have not as yet been found.

The gauge vector bosons of the Generation Group also have large masses. If the conservation of the fermion particle numbers is broken then we view it as a consequence of Generation Group symmetry breaking.

3-A.2 Basis of the Layer Group

The set of particle number operators can be further refined if we take account of the fourfold fermion generations. To further refine the set of particle number operators we temporarily neglect all interactions that would violate conservation laws for the set.

We therefore subdivide the above particle number set into four particle numbers per generation. For the ith generation we define

L_{ie} – The "e" species particle number for the ith generation

---

[16] By analogy, we assume that there are four species of Dark matter: charged Dark leptons, neutral Dark leptons, Dark up-type quarks, and Dark down-type quarks. Thus we are led to the Dark particle numbers: Dark Baryon Numbers, and Dark Lepton Numbers shown above.

L_{iv} – The v species particle number for the ith generation
L_{iuq} – The up-quark species particle number for the ith generation
L_{idq} – The down-quark species particle number for the ith generation

L_{iDe} – The Dark "e" species particle number for the ith generation
L_{iDv} – The Dark v species particle number for the ith generation
L_{iDuq} – The Dark up-quark species particle number for the ith generation
L_{iDdq} – Dark down-quark species particle number for the ith generation

for each generation i = 1, 2, 3, 4. Individual fermions have positive L_{ia} = +1 values and anti-fermions have negative L_{ia} = –1 values for species a = 1, 2, 3, 4 (with the three color subspecies of quarks treated as part of one species.)

At this point we have four particle number operators for each generation. We define a group framework for each set of particle numbers. The simplest way is to assume that each generation consists of four layers with the particles in each generation in a U(4) fundamental representation.[17] Then each generation has a U(4) Layer group with the generation's four number operators (above) as its diagonal operators. We call this group the Layer Group of the ith generation L_{ia}. With four generations we obtain four U(4) Layer groups for normal matter. In addition there are four U(4) Dark Layer groups. See Fig. 2.4.

The consequence of this expansion of particle numbers and groups is that the set of fermions increases fourfold. We now have four layers, with each having four generations, Experimentally, we know of three generations of fermions—the lowest generations of the lowest level. The remaining generation and three levels of fermions are of much higher mass and yet to be found.

See Blaha (2019g) and (2018e) for a detailed discussion of the Layer Groups. We note in passing that the symmetries of these number operators are badly broken. Yet the underlying group structure remains.

---

[17] See Fig. 2.3 for a depiction of the "splitting" of fermions: first into generations, then into layers.

4. Quaternion Universe–Octonion Megaverse Scenario

In earlier books[18] this author has shown that distortions in the structure of the universe suggest the existence of entities outside the universe. A space containing univereses was suggested that we called the Megaverse to distinguish it from the various multiverse theories. The contents of the Megaverse were described in some detail such as estimates of the size and average separation of constituent universes. The dimension of the Megaverse and its metric structure were hypothesized.

We now suggest structures for our universe, and for the Megaverse, that arise from a correspondence between the subgroups of the Complex Lorentz group and the symmetries of the Standard Model: U(1), SU(2) and SU(3).

In Blaha (2019f) and earlier books the symmetries of the Complex Lorentz group, as evidenced by their transformations, showed a close resemblance to the internal symmetries and transformations of the Standard Model.

Consequently the possibility arose of a larger space that supported complex coordinates with Complex Lorentz group transformations in one sector, and internal symmetries in another sector. This possibility was the source of the author's efforts here and in Blaha (2020). A consideration of the fermion particle spectrum—particularly the four species of fermions created by Complex Lorentz boosts—led to the consideration of a hypercomplex coordinate system for the universe with the feature of having a quaternion time coordinate and a set of three spatial biquaternions. (Similar considerations occurred for the Megaverse using a complex octonion space.)

The full space was a complex quaternion 4-space. Again following the lead of the Complex Lorentz group and Standard Model symmetries of Complex space-time, a complex quaternion 4-space was indicated. This space-time was assumed to mirror the

---

[18] See Blaha (2017c), (2018e),and (2019f) for detailed discussions.

complex space-time of our experience by supporting a complex quaternion Lorentz group.

As we shall see we found that a complex space-time sector, and an internal symmetry sector, existed with the symmetries of the Unified SuperStandard Theory. Thus a much deeper, coherent theory emerged that yielded the Unified SuperStandard Theory.

From the 4-dimensional complex quaternion space-time we then developed a larger theory for the Megaverse. This theory was based on an 8-dimensional complex octonion space. This space again "factored" to an 8-dimensional complex-valued space-time, and a set of internal symmetries, that included those of the 4-dimensional universe complex quaternion theory and added further symmetries that constituted an extension of the complex quaternion symmetries set. See chapters 8 - 11 for the details.

Thus a coherent scenario emerged of universes such as our universe embedded in a larger space.

4.1 Pattern of Fermion Particle Spectrum and Internal Symmetries of Standard Model and Unified SuperStandard Theory

Our work on the Unified SuperStandard Theory, and in previous books, has uncovered a close similarity between the internal symmetries of the Standard Model sector and subgroups of the Lorentz group. They both exhibit U(1), SU(2) and SU(3) symmetries.

Fermion Species
In addition we showed Lorentz group boosts have four varieties that map nicely to the four types of species of the fundamental fermions: charged lepton, neutral lepton, up-type quarks and down-type quarks IF boosts are restricted to real time and energy.

Parallel Internal Symmetry and Complex Lorentz Subgroups
The similarities and success in understanding the origin of the four species of fundamental fermions led us to consider the possibility that a wider space might encompass both space-time and the internal symmetry groups in a manner that does not violate known "NoGo" theorems.

In Blaha (2019e) and earlier books we showed a close parallel for U(1)⊗SU(2) and SU(3). Again the use of real-valued time was important. See section 6.4 with the explicit use of real-valued time and complex-valued spatial coordinates in the case of quaternions.

4.2 Hypercomplex Number Based Higher Dimensions

The Unified SuperStandard Theory was based on complex-valued coordinates. In choosing a higher dimension space for a larger theory of elementary particles the use of coordinate systems based on hypercomplex number systems seemed reasonable. The pattern of rising hypercomplexity is:

Real → Complex → Quaternion → Biquaternion → Octonion → Bioctonion

The Unified SuperStandard Theory took particle theory from real-valued coordinates to complex-valued coordinates. Complex quaternion (biquaternion) and complex octonion (bioctonion) extensions took us to QUeST and MOST that used larger spaces to unite space-time symmetry and internal symmetry.

Since the requirement of parallel physical processes made the minimal space-time dimension 4 and since the Megaverse must include universes as subspaces, we were led to a 4-dimensional complex quaternion formulation for our universe and an 8-dimensional complex octonion formulation for the Megaverse.

An intermediate possibility a 6-dimensional coordinate system combining a quaternion plus an octonion based coordinate system does not give internal symmetries of the Standard Model and Unified SuperStandard Theory

See Fig. 4.1 below for dimension and particle number comparisons for a complex quaternion universe (QUeST) and a complex octonion Megaverse (MOST).

Space	Complex Quaternion[19]	Complex Octonion[20]
Number of Quaternions[21] per Space-Time Coordinate	2	4
Number of Quaternion/Octonion Space-Time Dimensions	3 + 1	7 + 1
Total Number of Dimensions (real)	64	128
Internal Symmetry Dimensions (real)	52	104
Space-Time Dimensions (real)	8	16
Number of Fermions Per Layer[22]	64	128
Total Number of Fermions	256	512

Table 4.1. Comparison of QUeST universe and MOST Megaverse Features.

4.3 Glossary of Table 4.1 Entries

Number of Quaternions – The dimension of a coordinate measured in multiples of four..
:

---

[19] From chapter 6.

[20] From chapter 9.

[21] The 3 quaternion case is not considered here. It seems unsatisfactory because it does not lead to the desired internal symmetries or fermion spectrum. Cases where the number of quaternions exceed 4 appear to have much too many internal symmetries and particle spectra.

[22] The fermion number counts quark triplets: three types for each quark species.

Number of Space-Time Dimensions – The number of time and spatial quaternion or octonion coordinates.

Total Number of Dimensions – The total number of coordinates counting all time and spatial coordinates within quaternions (octonions).

Internal Symmetry Dimensions – The number of dimensions devoted to being internal symmetry representation coordinates.

Space-time Dimensions – The number of coordinates that become space-time coordinates.

Number of Fermions per Layer – The number of fundamental fermions in each layer counting individual quarks in quark triplets.

Number of Fermions – The sum over layers of the fundamental fermions counting individual quarks in quark triplets.

5. Quaternion Unified SuperStandard Theory (QUeST)

Quaternions are hypercomplex numbers that furnish a framework for defining higher dimension coordinate systems. Quaternions have significant properties that distinguish them:

1 .They are associative.

2. They are one of the two finite dimensional division rings having the real numbers as a proper subring. (The other is octonions—considered in chapter 8.)

3. They are non-commutative. (This is not a roadblock for quantum field theory which is also non-commutative in general.)

These features support the development of physics theories.[23]

5.1 Some Basic Quaternion Features

A quaternion is a 4-tuple of real numbers. A complex quaternion is a 4-tuple of complex numbers:

$$x = a + bi + jc + kd \ = a + \mathbf{v} \tag{5.1}$$

where a, b, c, d are real or complex numbers, and $\mathbf{v}$ is a 3-vector. The symbols i, j, and k are fundamental quaternion units. A quaternion norm is defined by

---

[23] There is an extensive literature on quaternions starting with the original work of Hamilton. Some recent, relevant papers are: S. L. Adler, "Generalized Quantum Dynamics", IASSNS –HEP-93/32 (1993); S. De Leo, arXiv:hep-th/9506179 (1995); Rolf Dahm, arXiv:hep-th/9601207 (1996); S. De Leo, arXiv:hep-th/9508011 (1995); S. L. Adler, arXiv:hep-th/9607008 (1996) and references therein.

$$\|\mathbf{x}\| = \text{sqrt}(aa^* + bb^* + cc^* + dd^*) \tag{5.2}$$

and the norm of **v** is

$$\|\mathbf{v}\| = \text{sqrt}(bb^* + cc^* + dd^*) \tag{5.3}$$

An important identity is

$$e^x = e^a (\cos(\|\mathbf{v}\|) + \mathbf{v}/\|\mathbf{v}\| \sin(\|\mathbf{v}\|))s \tag{5.4}$$

.It can be used to define boosts in quaternion space.

5.2 Motivation and Procedure

Our goal is to create a larger dimension space within which we can derive our space-time and the Unified Superstandard Theory in such a way as to understand the similarity of Lorentz subgroups and Standard Model internal symmetry groups. The development of a deeper basis for the Unified SuperStandard Theory will lead to refinements in the theory.

There are two possible procedures to follow in developing the deeper basis:

1. One can develop the Quantum Mechanics and Quantum Field Theory in a quaternion space and then extract the dynamics, fermion spectrum, gauge fields, and so on of our familiar space-time.

2. One can define a quaternion space, and then using its coordinates, directly extract the space-time, internal symmetries, fermion spectrum, gauge field spectrum and dynamics. Quaternion algebra may not be used.

We have chosen the latter approach as it will directly lead to the Unified SuperStandard Theory.

In developing the deeper space, upon which we will build, we will take guidance from the derivation of the Unified SuperStandard Theory. That theory assumes a complex 4-dimensional space-time upon which Complex General Relativity is constructed. It then proceeds to complex flat space-time and Complex Relativity.

We follow the procedure described in chapter 4 emulating the approach of the 4-dimensional Unified SuperStandard Theory.

5.3 Definition of Quaternion Space – Complex Quaternion (Biquaternion) Space

Following the stated procedure we define a complex or biquaternion space[24] with one "time" biquaternion and three "spatial" biquaternions believing the 3+1 space-time of our experience is a reflection of this deeper level.

Time Biquaternion
$$t = (a + bi + jc + kd) + I(a' + b'i' + j'c' + k'd')$$

Spatial Biquaternions
$$x = (a_x + b_x i + jc_x + kd_x) + I(a'_x + b_x'i' + j'c_x' + k'd_x')$$
$$y = (a_y + b_y i + jc_y + kd_y) + I(a'_y + b_y'i' + j'c_y' + k'd_y')$$
$$z = (a_z + b_z i + jc_z + kd_z) + I(a'_z + b_z'i' + j'c_z' + k'd_z')$$

where I is another "imaginary" number with $I^2 = -1$ that explicitly *makes biquaternions into complex quaternions. We will not use the algebra of quaternions but simply treat the quaternion coordinates as coordinates in a space.*

Fig 5.1 symbolically depicts the space with a black circle for each real-valued coordinate. We initially treat the real-valued and imaginary parts of a complex quaternion as a set of all real-valued coordinates.

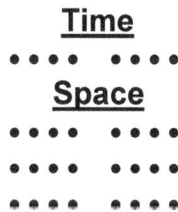

Time

•••• ••••

Space

•••• ••••

•••• ••••

•••• ••••

Figure 5.1. Four-Dimensional biquaternion space with coordinates represented by • 's.   There are 32 real-valued coordinates.

---

[24] We will use biquaternion to mean complex quaternions from this point on.

This biquaternion space has 32 real dimensions (16 complex dimensions.).

5.4 Biquaternion Lorentz Group

Our definition of time and space biquaternion coordinates purposefully resembles those of our real space-time. One might ask why should there be a Lorentz-like group for biquaternion space. The only apparent reason is the need for a special speed c in our space-time that enables one to boost from a rest frame of a mass m particle with energy m to a moving frame of greater energy.[25] Without c the group of the above coordinates would presumably be the biquaternion U(4) group. In this group transformations preserve the norm of a state so that a "boost-like" transformation does not exist.

The biquaternion Lorentz transformations do have a unique speed c (the speed of light) and specify a unique rest frame for any particle—both sublight particles and tachyon particles.[26] . Thus we select the Biquaternionic Lorentz group for biquaternion space.

Flat space biquaternion Special Relativity generalizes directly to a biquaternion General Relativity which may be constructed directly (mindful of quaternion non-commutativity).

The flat space biquaternion Lorentz group transformations have constant biquaternion matrix elements that are analogous to those of the Lorenz group.. (See Appendix A for a Lorentz group boost.)

5.5 Extracting the Symmetries and Particle Spectra

As stated earlier in section 5,2 we will directly describe the symmetry structure implied by the form of the biquaternion coordinate system while mindful of sections 5.3 and 5.4.

The Unified SuperStandard Theory developed the group structure from which the particle species were derived from a subset of Lorentz boost transformations. It used

---

[25] A similar consideration applies to tachyons.
[26] Tachyons can be transformed by a Complex Lorentz transformation to and from a rest frame. See Blaha (2018f) and Appendix A below.

boosts with complex exponentiation similar to quaternion exponentiation in eq. 5.4. Complex boosts mapped a system at rest to a system in motion with a real energy and complex 3-momenta in general. Biquaternion boosts play a similar role. See chapter 4.

The 4-dimensional representation of the Unified SuperStandard Theory complex coordinates was

<u>**Time**</u>

•

<u>**Space**</u>

• •

• •

• •

Figure 5.2. Four-Dimensional subspace for Unified SuperStandard derivation of particle spectra with coordinates represented by • 's.

Following the same line of reasoning we now specify a biquaternion space restricted to that of Fig. 5.3 to define the relevant set of coordinates for determining particle symmetries and spectra. Note time is a "real" quaternion. Spatial coordinates consist of complex quaternions.

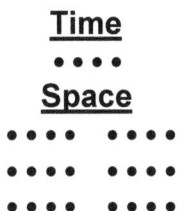

<u>**Time**</u>

• • • •

<u>**Space**</u>

• • • •   • • • •

• • • •   • • • •

• • • •   • • • •

Figure 5.3. 4-dimensional biquaternion subspace for symmetries and particle spectra.

We expect its 14 complex coordinates will split into a 4-dimensional complex coordinate space[27] which will support Complex Lorentz transformations, and 10 complex coordinates for internal symmetry space. Internal symmetry space has an initial SU(10) group before breakdown.

The mechanism for this symmetry breakdown may involve vacuum energy effects in the biquaternion universe.

In the next chapter we analyze the symmetries of the 10 complex dimensional subspace and then augment these internal symmetries with Generation and Layer number symmetries.

In MOST (chapters 8 and 9) we show the number symmetries are inherently part of the set of internal symmetries. We also show that Dark fermion sectors of the theory have spinors that occupy different parts of the overall 16 component spinors of the complex 8-dimensional space-time.

These MOST features will feed back to a deeper version of QUeST that does not disagree with features stated in this chapter and chapter 6.

---

[27] The 3+1 biquaternion space becomes a 4-dimensional complex coordinate space which then becomes the 4-dimensional real-valued space of our experience.

6. Symmetries of the Quaternion Unified SuperStandard Theory (QUeST)

The coordinate space picture of QUeST described in chapter 5 enables us to simply find the internal symmetries and particle spectra of QUeST. They will turn out to be those of the Unified SuperStandard Theory as presented earlier in chapter 2.

The QUeST coordinate subspace for the determination of internal symmetries is depicted in Fig. 6.1..It is based on the discussion of Fig. 5.3. The complex space-time 4-vector is separated from the internal symmetry coordinates.

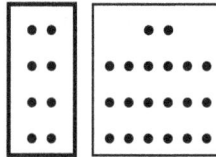

Figure 6.1. Internal Symmetry determination subspace. Eight real-valued space-time coordinates (complex 4-vector) are separated from 20 real-valued coordinates for internal symmetries.

The internal symmetry coordinates number 20 real coordinates or 10 complex coordinates. These coordinates serve as coordinates[28] of the fundamental representations of each of the factors of

$$SU(2)\otimes U(1)\otimes SU(3)\otimes SU(2)\otimes U(1)\otimes SU(3) \tag{6.1}$$

---

[28] Simple counting of fundamental representation dimensions shows this to be true:: $2 + 3 + 2 + 3 = 10$ respectively. The set of 10 complex coordinates support transformations to a factorized block-diagonal form. The 10 complex coordinates can be transformed into a fundamental representation of the above factor groups.

The factorized internal symmetry emerges from another breakdown(s) which corresponds to the subgroups of the Lorentz group. They evidently follow from the structure of biquaternion Lorentz transformations.

6.1 Number Symmetries

The U(4) Generation and Layer groups do not appear in Fig. 6.1 or eq. 6.1. Additional quaternion coordinates must be added to represent them. Four quaternions suffice to represent :for Generation groups a U(4) for the normal sector and a U(4) for the Dark sector; and for the Layer groups a U(4) for the normal sector and a U(4) for the Dark sector. Fig. 6.2 illustrates the coordinates for $U(4)^4$.to be added to the dimensions in Fig. 5.1.

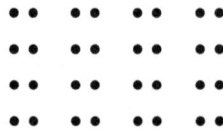

$$
\begin{array}{cccc}
\cdot\cdot \quad \cdot\cdot & \cdot\cdot & \cdot\cdot \\
\cdot\cdot \quad \cdot\cdot & \cdot\cdot & \cdot\cdot \\
\cdot\cdot \quad \cdot\cdot & \cdot\cdot & \cdot\cdot \\
\cdot\cdot \quad \cdot\cdot & \cdot\cdot & \cdot\cdot
\end{array}
$$

Figure 6.2. The four addional biquaternions needed for normal and Dark Generation and Layer groups $U(4)^4$.

Later when we discuss MOST we will see that these additional quaternion coordinates arise naturally when the MOST group structure is projected back to the QUeST sector.

We will also see that the total additional dimensions lead to particle-coordinate duality. (See Chapter 7.)

6.2 Layers

The QUeST figures above do not display the Internal Symmetry, Generation and Layer groups in four layers. The groups differ in the various layers. Their gauge fields are each flagged with a layer index. So the overall internal symmetry is the internal symmetry group of the Unified SuperStandard Theory:

$$[SU(2)\otimes U(1)\otimes SU(3)\otimes SU(2)\otimes U(1)\otimes SU(3)\otimes U(4)^4]^4 \qquad (6.2)$$

6.3 Factorization of the Internal Symmetries of a Layer

The 10 plus 16 additional coordinates totaling 26 complex coordinates form an SU(26) internal symmetry space. This space undergoes a breakdown. The resulting factorized form of the internal symmetries *of one layer*[29]   is

$$SU(2) \otimes U(1) \otimes SU(3) \otimes SU(2) \otimes U(1) \otimes SU(3) \otimes U(4)^4 \qquad (6.3)$$

This is represented by Fig. 6.3. The seemingly duplicate factors of eqs. 6.2 and 6.3 are for the "normal" and the Dark sectors. They are needed to avoid interactions that would "ruin" the Darkness of the Dark sector.

Dark1: U(1)⊗SU(2)⊗SU(3)   Dark2: U(1)⊗SU(2)⊗SU(3)

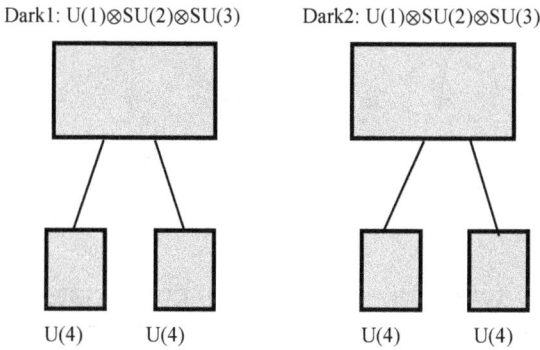

U(4)   U(4)   U(4)   U(4)

Figure 6.3. Schematic of the internal symmetry groups of eq. 6.3. This figure combines Figs. 6.1 and 6.2. The two large blocks are each 10 real coordinate representations of SU(2)⊗U(1)⊗SU(3). The lower U(4) groups are the Generation and Layer number groups. One pair of each number group is for each of the two U(1)⊗SU(2)⊗SU(3) factors above.

---

[29] Strictly speaking this is a simplification since the Layer group mixes all layers.

6.4 The Breakdown of the Symmetry of the 20 Real Coordinate Group SU(10)

The 20 real coordinate subspace of Fig. 6.1 contains a representation of

$$SU(2) \otimes U(1) \otimes SU(3) \otimes SU(2) \otimes U(1) \otimes SU(3) \tag{6.4}$$

which clearly could be transformed to a block diagonal in the group factors.

However it is instructive to consider the generation of the component factors using a Lorentz boost-like framework. Separating the internal symmetry part of Fig. 6.1 into six sets of Lorentz-like coordinates with a real energy and complex momenta we obtain Fig. 6.4.

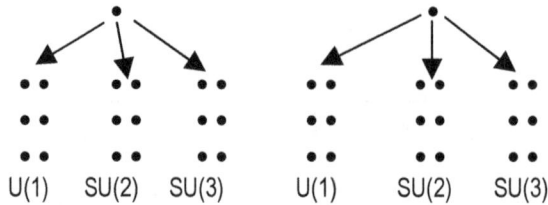

Figure 6.4. Each of the two real-valued time coordinate is linked to three complex spatial coordinates. Lorentz transformations applied to each 4-vector so constructed yields a factor of eq. 6.4. The factors so generated commute to give eq. 6.4 with commuting subgroup factors listed in the figure.

The use of coordinates with real-valued energy and complex-valued spatial momentum was shown in chapter 4 to lead to the four fermion species using Lorentz boosts. Here we see a similar effect in the generation of particle interactions.

6.5 Map to Unified SuperStandard Theory Groups

The map of the four complex coordinates to Complex space-time is direct. Complex Lorentz group transformations relate coordinate systems.

The complete set of internal symmetries for one layer is given by eq. 6.3. The map follows from the observations:

1. The factors of eq. 6.3 can be separated into factors for normal and for Dark sectors.

2. Since the set of normal and Dark fermions are split into four species we can unambiguously associate the factors of eq. 6.3 with its factors.

3. The SU(3) factors have $\underline{3}$ representations which we can associate with the up-quark and down-quark, normal and Dark species. Thus the SU(3) subgroups are normal and Dark Strong interaction subgroups.

4. The SU(2)⊗U(1) factors map to ElectroWeak interactions for the normal and Dark sectors.

5. The four U(4) factors map to the Generation and Layer number groups for the normal and for the Dark sectors.

Thus we have a map to the interactions of the Unified SuperStandard Theory. Although we consider only one layer the considerations apply to all four layers.

6.6 Fermion and Gauge Vector Boson Spectrums

The resulting fermion and vector boson spectrums that emerge are those of the Unified SuperStandard Theory. They are displayed in Figs. 2.1 – 2.4.

7. Particle-Dimension Duality

An examination of the total number of dimensions and the number of fundamental fermions (per layer) as we saw in Table 4.1 shows that these quantities are equal for the QUeST case (64) and the MOST case (128) seen later. The equality of these parameters raises the possibility that the number of internal symmetry dimensions equals the number of fundamental fermions in some deep sense.

When one considers the measurement of distance in a space one sees the measuring process requires the use of particles. Without particles one cannot distinguish space-time points from each other. And the distance between space-time points is determined either directly or indirectly by bosons such as light transmitted (and possibly reflected) between fermion "clumps.'

Without particles distance, space-time is meaningless. Without space-time particles will "clump' and dynamics is impossible.

Thus a relation between particles and dimensions is required for Physics.

7.1 Particles Mapped to Quantum Functionals

A particle has a quantum field. We have shown in earlier books that quantum fields may be factored into an inner product of a quantum functional and a wave in coordinate space-time.[30] Every particle has a set of internal symmetry and space-time quantum numbers. We can map the set of fundamental fermions to a set of quantum functionals. These functionals have transformation properties under internal symmetries. They are thus representations of the internal symmetry groups.

There is no distance in the set of functionals. As we showed in Blaha (2020) and earlier books factoring quantum fields into inner products of functionals and waves eliminates the instantaneity problem of Quantum Entanglement.

---

[30] See Blaha (2019e) and (2020) as well as earlier books by the author.

7.2 Coordinates and Functionals

The lack of distance in the set of functionals parallels the lack of distance between coordinates. For example there is no distance between the "x" coordinate and the "y" coordinate in a coordinate system. Functionals have features in common with coordinates.

Quantum functionals can also have commutation relations similar to coordinates: Defining a quantum functional conjugate momentum:

$$\pi_k = d/df_k \tag{7.1}$$

For quantum functional f_k we obtain the commutation relation

$$[f_k, \pi_j] = i\delta_{kj} \tag{7.2}$$

where k and j represent internal symmetry indices. Eq. 7.2 mirrors the form of quantum mechanical commutation relations adding to the similarity between coordinates (dimensions) and quantum functionals.

Biquaternion and bioctonion coordinates can be mapped to quantum functionals since they have the same role, and number, as the coordinates of internal symmetries. We set

$$x_m = R_{mn}f_n \tag{7.3}$$

where R is a transformation.

7.3 Particle-Dimension Duality

Thus in QUeST and MOST we find a particle-dimension (coordinate) duality.[31] The point of view offered by this duality suggests that fermion particles are in some sense "interchangeable" with internal symmetry coordinates—with particle functionals as an intermediary.

---

[31] Although the discussion is for one layer of particles it can be generalized to four layers of fermions with the introduction of additional sets of coordinates.

7.4 Particle – Functional – Dimension Triality

The above discussion shows a close analogy – a *triality* – between fermion particles, dimensions (coordinates) and quantum functionals. We will pursue these topics in more detail elsewhere.

8. Bioctonion Megaverse

In chapters 5 and 6 we developed a biquaternion theory called QUeST that could be used to derive the group structure, and the fundamental fermion and vector boson spectrums. It led to the Unified SuperStandard Theory and accounted for the close similarity between the internal symmetries of the Standard Model sector and the subgroups of the Lorentz group. They both exhibit U(1), SU(2) and SU(3) symmetries.

In this chapter we define a bioctonion space and use it to define a Megaverse fundamental basis of a somewhat more general Unified SuperStandard Theory, called MOST. MOST develops a more robust set of internal symmetries and fundamental particles. It creates a new view of Dark matter that appears to help explicate the lack of interactions between normal matter and Dark matter.

In chapters 5 and 6 we found the Unified SuperStandard Theory, which was based on the Complex Lorentz group, could be based on a biquaternion space for our universe.

If we assume the existence of a Megaverse[32] containing our universe, and other universes, then we can define a *Megaverse Octonion SuperStandard Theory* (MOST) that becomes a more general basis of the Unified SuperStandard Theory.

Remarkably MOST, when "restricted" to our universe, yields QUeST with the number symmetries that we added "by hand" to QUeST. It also suggests that at least part of the Darkness of Dark matter may be due to differing spinors for Dark fermions that inhibit matter – Dark matter interactions.

8.1 Octonion Features

Octonions have significant properties that enable them to be used in a quantum field theory development:

---

[32] The Megaverse is described in some detail in Blaha (2017c), (2017f), and (2018e) together with evidence for its existence.

1. An octonion is an 8-tuple of real numbers. A complex octonion is an 8-tuple of complex numbers.
2. They are nonassociative.
3. They are one of the two finite dimensional division rings having the real numbers as a proper subring. (The other is quaternions—considered in chapter 5.)
4. They are non-commutative. (This is not a roadblock for quantum field theory which is also non-commutative in general.)

These features support the development of physics theories.
We can represent a bioctonion b as

$$b = b_{real} + I b_{imaginary}$$

where b_{real} and $b_{imaginary}$ are real-valued octonions.

8.2 Motivation and Procedure

Our goal again is to create a larger dimension space within which a universe can exist based on QUeST, and where we can ultimately derive our space-time and the Unified Superstandard Theory. Again we use the similarity of Lorentz subgroups and Standard Model internal symmetry groups in our development.
There are two possible procedures to follow in developing the deeper basis:

1. One can develop the Quantum Mechanics, Quantum Field Theory, … in an octonion space and then extract the dynamics, fermion spectrum, gauge fields, and so on of our familiar space-time.

2. One can define an octonion space and then use its coordinates to directly extract the space-time, internal symmetries, fermion spectrum, gauge field spectrum and dynamics.

We have chosen the latter approach as it will more directly lead to the Unified SuperStandard Theory.

In developing the deeper space, upon which we build, we will take guidance from the derivation of the Unified SuperStandard Theory. This theory assumes a complex 4-dimensional space-time upon which Complex General Relativity is constructed. It then proceeds to complex flat space-time and Complex Relativity.

After defining features of Complex Lorentz transformations the Unified SuperStandard Theory used Lorentz boosts to derive the Dirac forms of the four fermion species. The boosts were required to boost a fermion from a rest state to a state with a real-valued energy, and real or complex-valued 3-momenta. Thus a real time – complex-valued spatial part is required for the proper definition of species.

The Unified SuperStandard Theory then showed Lorenz subgroups mapped to Standard Model internal symmetry subgroups. See chapter 4 for a more detailed discussion.

8.3 Definition of Bioctonion Space

Following the above stated procedure we define a bioctonion[33] (complex octonion) space with *one* "time" biquaternion and *seven* "spatial" bioctonions as a generalization of the 3+1 space-time of our experience. The choice of 8 bioctonion dimensions seemed natural but was not required by a principle. It does lead to a larger set of internal symmetries with QUeST symmetries as a subset . We will use the symbol • to represent each of the 16 bioctonion coordinates in Fig. 8.1.

We have chosen a complex 8-dimensional bioctonion space-time as the Megaverse space-time. There are 128 real coordinates in the bioctonion "parent" space from which complex 8-dimensional space-time (4-dimensional complex space-time) is extracted. It embeds our universe's 4-dimensional complex space-time as a subspace-time.

Fig 8.1 symbolically depicts the space with a circle for each real-valued coordinate. *Again we treat the bioctonion space as a higher dimensional space and do not use details of octonion algebra in our development.*

---

[33] We use bioctonion synonymously with complex octonion in this and subsequent chapters.

Time

• • • • • • • •   • • • • • • •

7-Space

• • • • • • • •   • • • • • • •
• • • • • • • •   • • • • • • •
• • • • • • • •   • • • • • • •
• • • • • • • •   • • • • • • •
• • • • • • • •   • • • • • • •
• • • • • • • •   • • • • • • •
• • • • • • • •   • • • • • • •

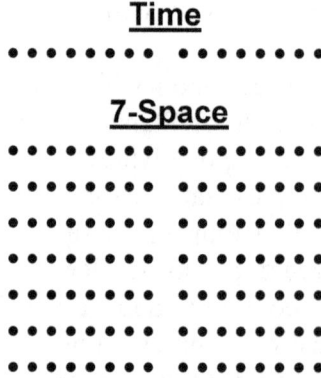

Figure 8.1. Eight-Dimensional (7 + 1) bioctonion space with coordinates represented by • 's.

This bioctonion space has 128 real dimensions (64 complex dimensions.).

8.4 Bioctonion Lorentz Group

Our definition of time and space bioctonion coordinates purposefully resembles those of our real space-time. One might ask why there should be a Lorentz-like group for bioctonion space. up.

The only apparent reason is the need for a special speed c that, in our space-time, its existence enables one to boost from a rest frame of a mass m particle with energy m to a moving frame of greater energy.[34] Without c the group of the above coordinates would presumably be the bioctonion U(8) group. U(8) transformations preserve the norm of a state so that a "boost-like" transformation does not exist.

The bioctonion Lorentz transformations do have a unique speed c (the speed of light) and specify a unique rest frame for any particle—both sublight particles and tachyon particles.[35] . Thus we select the bioctonion Lorentz group for bioctonion space.

---

[34] A similar consideration applies to tachyons.
[35] Tachyons can be transformed by a Complex Lorentz transformation to and from a rest frame. See Blaha (2018f) and Appendix A below.

Flat space bioctonion Special Relativity generalizes to a bioctonion General Relativity which may be constructed directly (mindful of octonion non-commutativity).

Flat space bioctonion Lorentz group transformations have constant bioctonion matrix elements that are analogous to those of the Lorenz group.

8.5 Extracting the Symmetries and Particle Spectra

As stated earlier in section 8.2 we will directly describe the symmetry structure implied by the form of the bioctonion coordinate system while mindful of sections 8.3 and 8.4.

The Unified SuperStandard Theory developed the group structure, from which the particle species were derived, from a subset of Lorentz boost transformations. Complex boosts mapped a system at rest to a system in motion with a real energy and complex 3-momenta in general. Bioctonion boosts play a similar role.

The 4-dimensional representation of the Unified SuperStandard Theory complex coordinates is given in Fig. 8.2.

Time

•

Space

• •

• •

• •

Figure 8.2. Four-Dimensional space for Unified SuperStandard derivation of particle spectra with coordinates represented by • 's.

Following the same line of thought, which is described in more detail in chapter 4, we now specify a bioctonion subspace restricted to that of Fig. 8.3 to define the relevant set of coordinates for determining particle symmetries and spectra.

Time

• • • • • • • •

7-Space

• • • • • • • •   • • • • • • • •
• • • • • • • •   • • • • • • • •
• • • • • • • •   • • • • • • • •
• • • • • • • •   • • • • • • • •
• • • • • • • •   • • • • • • • •
• • • • • • • •   • • • • • • • •
• • • • • • • •   • • • • • • • •

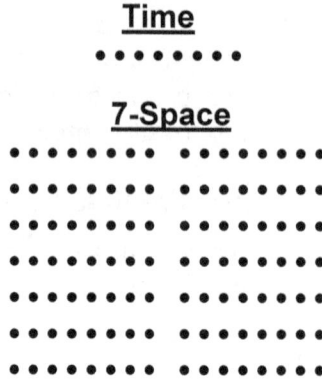

Figure 8.3. 8-dimensional bioctonion subspace for symmetries and particle spectra.

Its 60 complex coordinates will split into an 8-dimensional complex coordinate space which will support 8-dimensional Complex Lorentz transformations, and a 52 complex coordinates internal symmetry space.

The mechanism for this symmetry breakdown may be due to vacuum energy effects in the bioctonion Megaverse.

In the next chapter we analyze the symmetries of the 52 complex dimensional subspace. We find it contains the Standard Model symmetries and the Generation and Layer number symmetries. The number symmetries are inherently part of the set of internal symmetries in this case. We will also see later that the Dark fermion sectors of the theory have spinors that occupy different parts of the overall 16 component spinors of complex 8-dimensional space-time.

9. Symmetries of the Megaverse Octonion SuperStandard Theory (MOST)

The coordinate space picture of the Megaverse described in chapter 8 enables us to simply find the internal symmetries and particle spectra of MOST. They will turn out to be a superset of those of the Unified SuperStandard Theory as presented earlier in chapter 2.

The MOST Megaverse subspace for the determination of the Internal Symmetries is depicted in Fig. 9.1..It is based on the discussion of Fig. 8.3. The space-time complex 8-vector is separated from the internal symmetry coordinates.

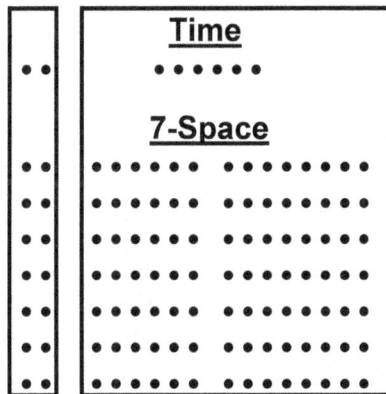

Figure 9.1. Internal Symmetry determination subspace. It is Based on Fig. 8.3; The 16 real space-time coordinates are separated from 104 real coordinates for internal symmetries.

The internal symmetry coordinates above number 104 real coordinates or 52 complex coordinates. These coordinates serve as the coordinates of the fundamental representations of each of the factors of

$$[SU(2) \otimes U(1) \otimes SU(3) \otimes SU(2) \otimes U(1) \otimes SU(3)]^2 \otimes U(4)^8 \qquad (9.1)$$

The factorized internal symmetry emerges from another breakdown(s) which corresponds to the subgroup structure of the Lorentz group. Eq. 9.1evidently follows from the structure of bioctonion Lorentz transformations.

The U(4) Generation and Layer groups are now represented in Fig. 9.1. We depict the pattern of symmetry implied by Fig. 9.1 and eq. 9.1 in Fig. 9.2 and 9.3.

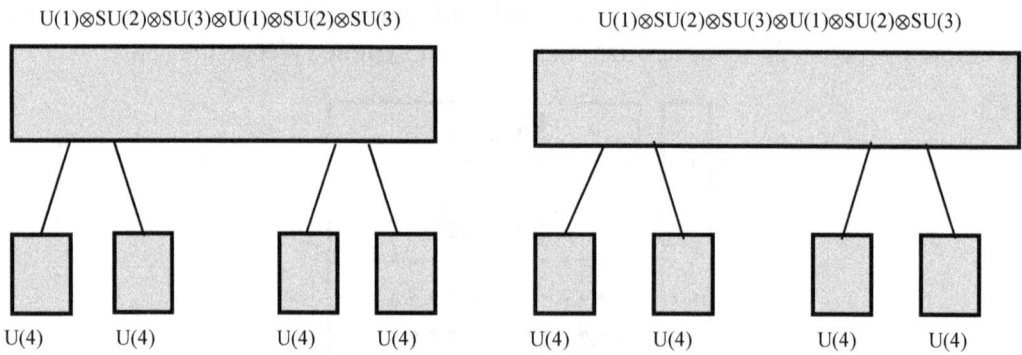

U(1)⊗SU(2)⊗SU(3)⊗U(1)⊗SU(2)⊗SU(3)   U(1)⊗SU(2)⊗SU(3)⊗U(1)⊗SU(2)⊗SU(3)

U(4)   U(4)   U(4)   U(4)   U(4)   U(4)   U(4)   U(4)

Figure 9.2. Schematic of the internal symmetry groups' coordinates of Fig. 9.1. The two "large" blocks are sets of 20 real-valued coordinates furnishing representations of the listed groups. The lower U(4) groups are the Generation and Layer number groups. The total number of real-valued coordinates is 104.

Each U(1)⊗SU(2)⊗SU(3)⊗U(1)⊗SU(2)⊗SU(3) block in Fig. 9.2 has a 10 complex coordinates (20 real-valued coordinates) representation. The blocks are subdivided in Fig. 9.3 into sets of 10 real-valued coordinates supporting representations of U(1)⊗SU(2)⊗SU(3). There are three Dark blocks. The first block contains the representations of the known parts of the Standard Model. The internal symmetry groups of each part are listed in Fig. 9.4.

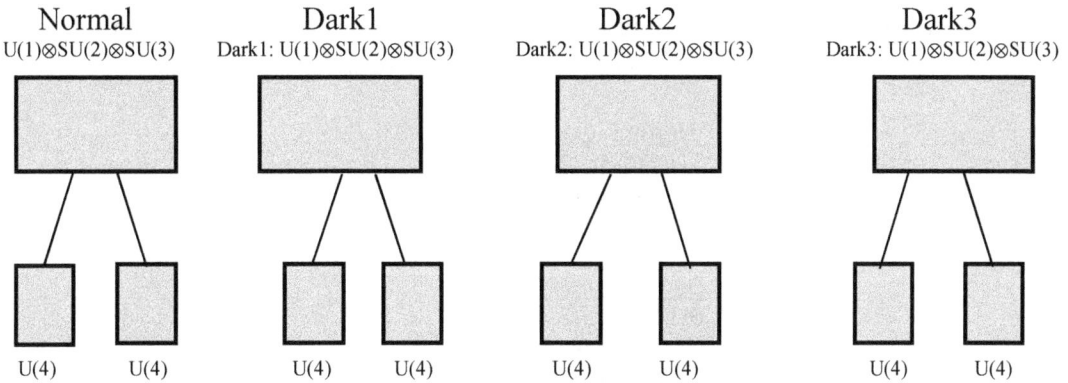

Figure 9.3. Schematic of the internal symmetry groups of eq. 9.1. The lower U(4) groups are the Generation and Layer number groups. One pair of each number group is for each of the four U(1)⊗SU(2)⊗SU(3) factors above. The result is the total Internal Symmetry group of the enlarged Unified SuperStandard Theory. (See chapter 2.)

The factorization of each of the four blocks is accomplished by following the procedure given in section 6.4 for each block.

9.1 Layers

QUeST and MOST give the internal symmetries of one layer. The symmetries of the three other layers are the same but their groups are different. Instead the groups of each layer are each flagged with a layer index.

So the overall internal symmetry is the internal symmetry group of QUeST and the Unified SuperStandard Theory plus two additional SU(2)⊗U(1)⊗SU(3) parts and their Generation and Layer groups:

$$[SU(2)\otimes U(1)\otimes SU(3)\otimes SU(2)\otimes U(1)\otimes SU(3)\otimes SU(2)\otimes U(1)\otimes SU(3)\otimes SU(2)\otimes U(1)\otimes SU(3)\otimes U(4)^8]^4 \quad (9.2)$$

9.2 Universe – Megaverse Connection

Comparing Figs.6.3 and 9.3 we see that a QUeST universe has a "Normal" part and one Dark part while the MOST Megaverse adds two more Dark parts.

Thus we can envision a universe being a subspace of the Megaverse with QUeST being a part of MOST. This is discussed in more detail in Chapter 12.

"Normal" Gauge Groups
SU(3)⊗SU(2)⊗U(1)
Generation Group U(4)
Layer Group U(4)

Dark1 Gauge Groups
SU(3)⊗SU(2)⊗U(1)
Generation Group U(4)
Layer Group U(4)

Dark2 Gauge Groups
SU(3)⊗SU(2)⊗U(1)
Generation Group U(4)
Layer Group U(4)

Dark3 Gauge Groups
SU(3)⊗SU(2)⊗U(1)
Generation Group U(4)
Layer Group U(4)

Figure 9.4. MOST vector bosons list from eq. 9.2. The four layers quadruple the above list: one distinct set for each layer.In one layer the total number of vector bosons of the above list is 176. Thus four layers yield a total count of 704 vector bosons in MOST.

9.3 Fermion and Gauge Vector Boson Spectrums

The fermion and vector boson spectrums that emerge in MOST are those of an enlarged QUeST and Unified SuperStandard Theory. They are displayed in Figs. 2.1 – 2.4..MOST has an additional two Dark sectors. (Figs. 9.3.and 9.4)

Vector Bosons

From Fig. 9.4 we find MOST has 176 vector bosons in one layer. Thus four layers yield a total count of 704 MOST vector bosons. There are two additional Dark vector boson sectors beyond QUeST and the Unified SuperStandard Theory.

Fermions

There are 512 fundamental fermions in MOST, which includes two additional Dark fermion sectors. Chapter 12 explains the transition from MOST to QUeST and the Unified SuperStandard Theory. Fig. 9.5 shows the MOST fermion spectrum.

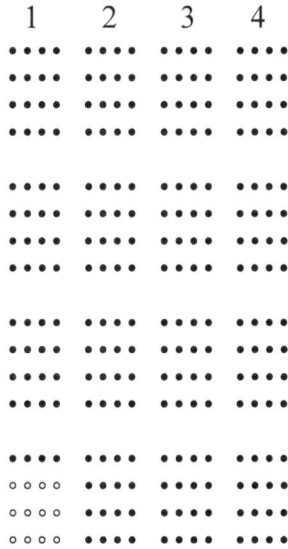

```
     1      2      3      4
   • •••  • •••  • •••  • •••
   • •••  • •••  • •••  • •••
   • •••  • •••  • •••  • •••
   • •••  • •••  • •••  • •••

   • •••  • •••  • •••  • •••
   • •••  • •••  • •••  • •••
   • •••  • •••  • •••  • •••
   • •••  • •••  • •••  • •••

   • •••  • •••  • •••  • •••
   • •••  • •••  • •••  • •••
   • •••  • •••  • •••  • •••
   • •••  • •••  • •••  • •••

   • •••  • •••  • •••  • •••
   ○ ○○○  • •••  • •••  • •••
   ○ ○○○  • •••  • •••  • •••
   ○ ○○○  • •••  • •••  • •••
```

Figure 9.5. Schematic spectrum of the fermions of MOST. Each fermion is represented by a •. Quark triplets are represented by a single •. Four sets of four species in four generations which are in turn in 4 layers. Open symbols ○ represent known fermions. There are 512 fundamental fermions taking account of quark triplets.

10. Darkness Superselection: Why Normal and Dark Particles Don't Interact

Despite numerous searches no interaction between normal and Dark matter has been found except for gravitation. Consequently a sizeable number of physicists has concluded that Dark matter does not exist.

Our study of MOST raises a contrary possibility:. A superselection rule for a conserved Darkness number that precludes all interactions between Dark and normal matter except gravitation. We suggest a quantum number, *Darkness* denoted D, exists, similar to charge, with integer values: 1 for normal matter, and 2, 3, and 4 for the Dark matter sectors of MOST. When telescoped downward to QUeST the Darkness quantum number D is 1 for normal matter and 2 for Dark matter. The columns in Fig. 10.1 are labeled with the MOST Darkness quantum number.

Interaction vector bosons also have the Darkness quantum number as shown in Fig. 10.2.

10.1 Darkness Bound Interactions

A superselection rule for D implies it is a conserved quantum number with no interactions between sectors with different values of D. Thus fermion – vector boson interactions have the form

$$\overline{\psi}_k A_k \psi_k \qquad (10.1)$$

for k = Darkness number.

No vector interactions exist between normal and Dark matter. D conservation implies the lack of matter – Dark matter interactions. Gravitation being universal does exist between all forms of matter. The gravitational spinor connection interaction can be defined to exclude interactions between D sectors.

The picture implied by D having a superselection rule appears to agree with experiment.

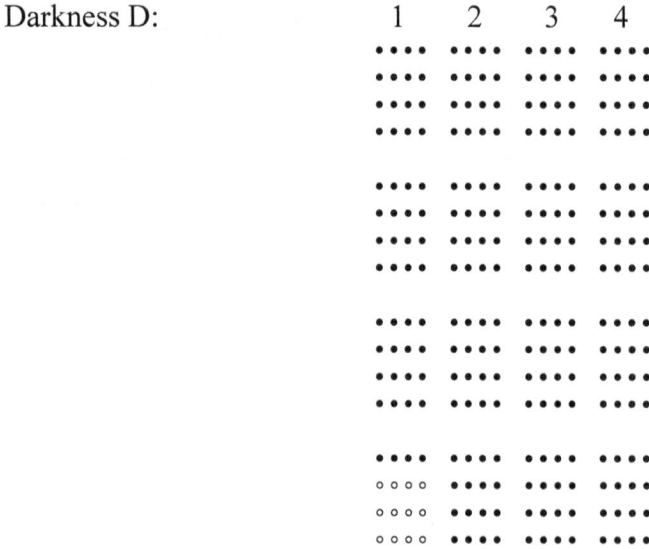

Darkness D:

```
                    1     2     3     4
                  • • • •  • • • •  • • • •  • • • •
                  • • • •  • • • •  • • • •  • • • •
                  • • • •  • • • •  • • • •  • • • •
                  • • • •  • • • •  • • • •  • • • •

                  • • • •  • • • •  • • • •  • • • •
                  • • • •  • • • •  • • • •  • • • •
                  • • • •  • • • •  • • • •  • • • •
                  • • • •  • • • •  • • • •  • • • •

                  • • • •  • • • •  • • • •  • • • •
                  • • • •  • • • •  • • • •  • • • •
                  • • • •  • • • •  • • • •  • • • •
                  • • • •  • • • •  • • • •  • • • •

                  • • • •  • • • •  • • • •  • • • •
                  ○ ○ ○ ○  • • • •  • • • •  • • • •
                  ○ ○ ○ ○  • • • •  • • • •  • • • •
                  ○ ○ ○ ○  • • • •  • • • •  • • • •
```

Figure 10.1. Schematic of the spectrum of fermion layers of MOST labeled with Darkness obtained from Fig. 9.5. The empty circles are known fermions. Quark triplets are represented by a ○ or a •. Darkness 1 includes the known fermions of the Standard Model. Darkness 1 and 2 are the fermions of the Unified SuperStandard Theory and QUeST in our universe. Darkness 1 through 4 appear in MOST in the Megaverse.

10.2 Other Superselection Rule Based Quantum Numbers

Each Darkness sector has its own charge and other internal symmetry quantum numbers. The Darkness superselection rule implies superselection rules for:

Charge
Other Internal Symmetry Quantum Numbers

SU(3) Numbers
SU(2)⊗U(1) Numbers
Generation Numbers
Layer Numbers
Spin
Total Angular Momentum

The spin superselection rule is required by conservation of angular momentum in each sector. See section 10.3.

The restriction of the above quantities to different Darkness sectors where they have their own specific conservation laws removes the special significance of Charge and other quantum numbers. They are not of universal significance. They are of sector-restricted significance.

The Darkness superselection rule, which precludes inter-sector interactions, implies the inter-penetrability of matte and energy. The absence of interactions enables matter and energy of all sectors to coincide at the same point; point by point, throughout the universe and Megaverse.

10.3 Darkness Structured Spinors

The case of the particle spin superselection rule is of importance in MOST and QUeST. First we note that total angular momentum must be separately conserved in each Darkness sector.

Secondly, we note eight-dimensional Megaverse space-time has 16 component Dirac fermion spinors. Given the split of the fermion spectrum into four sectors (Fig. 10.1) it is reasonable to decompose 16-spinors into four 4-dimensional Dirac spinors using spin projection operators Π_k.

$$
\begin{aligned}
\Pi_1 &= (I, 0, 0, 0) \\
\Pi_2 &= (0, I, 0, 0) \\
\Pi_3 &= (0, 0, I, 0) \\
\Pi_4 &= (0, 0, 0, I)
\end{aligned}
\tag{10.2}
$$

with fermion wave functions for each Darkness sector

$$\psi_i(x) = \Pi_i \psi(x) \tag{10.3}$$

using 4×4 matrices composed of 0's and the identity matrix I.

If we define all vector interactions to have a spin of the type of its Darkness sector then normal entities only interact with normal entities, and Dark entities of each sector only interact with other Dark entities in its sector. Total angular momentum is then conserved by interactions. (eq. 10.1)

Thus the absence of interactions between normal and Dark matter is enhanced.

Darkness D:	1	2	3	4
	Gauge Groups	Gauge Groups	Gauge Groups	Gauge Groups
	Gauge Groups	Gauge Groups	Gauge Groups	Gauge Groups
	Gauge Groups	Gauge Groups	Gauge Groups	Gauge Groups
	Gauge Groups	Gauge Groups	Gauge Groups	Gauge Groups
Generic Field:	A_1	A_2	A_3	A_4

Figure 10.2. The list of MOST vector bosons based on Fig. 9.4. There are four layers shown vertically. D = 1 includes the known vector bosons of the Standard Model. D = 1 and 2 are the vector bosons of the Unified SuperStandard Theory and QUeST in our universe. D = 1 through 4 appear in MOST in the Megaverse. The generic vector boson field for each value of D appears below the list.

10.4 QUeST 8-Spinors

In QUeST there is a norm matter sector with D = 1 and a Dark matter sector with D = 2. In chapters 5 and 6 we did not consider QUeST spin. We now suggest that spin "filters down" from MOST to QUeST. We choose to have 8-spinors with the upper four spinor components for normal $D = 1$ matter and the lower four components for Dark $D = 2$ matter.

Then using the above discussion restricted to $D = 1$ and $D = 2$ we obtain the scenario of section 10.1 and the Darkness superselection rule, and superselection rules for the quantum numbers in section 10.2, with the $D = 1$ and $D = 2$ Darkness sectors. The QUeST vector interactions have the form

$$\Sigma_k \ \bar{\psi}_k A_k \psi_k \tag{10.4}$$

for normal ($k = 1$) and Dark ($k = 2$) interactions.

Total angular momentum is separately conserved in each Darkness sector. Normal matter and Dark matter do not interact.

This discussion is directly used to generalize the Unified SuperStandard theory.

.

11. Megaverse-Universe Connection

MOST describes the unified elementary particle theory of the Megaverse. Universes within the Megaverse, which are subspaces of the Megaverse, are described by QUeST. Table 4.1 compares MOST and QUeST features.

QUeST is the 4-dimensional subset of MOST. As chapters 6 and 9 show QUeST has a subset of MOST fermions, vector bosons and interactions. MOST has four sectors of differing Darkness. The MOST 16-spinor, which is composed of four 4-spinors becomes an 8-spinor composed of two 4-spinors. The below figure depicts the relation between MOST and QUeST. (QUeST and the Unified SuperStandard Theory have the same features. They are both enhanced with some new features (shown in the dotted box) emanating from MOST as shown below.)

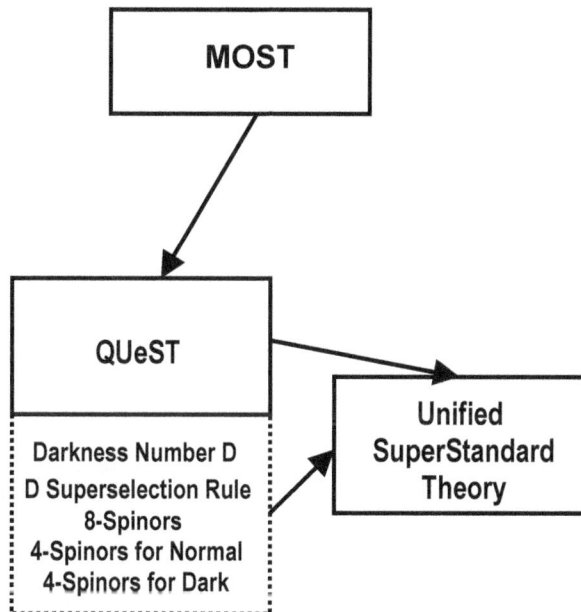

Appendix A. Fermion Species of QUeST, MOST, and the Unified SuperStandard Theory

In Blaha (2018e) (and earlier books) we showed that the overall form of the fundamental fermion spectrum consisted of four species that were determined by the restriction of Complex Lorentz boosts to boosts from a particle's rest frame to a coordinate system with a real-valued energy and time. There are four types of these boosts: two yield a normal particle and two yield a tachyon. Each of the two boosts of each type boost to a coordinate system with real-valued spatial coordinates and to a coordinate system with complex-valued spatial coordinates.

We find that this general form of the four boosts, and fermion species, occurs in the Unified SuperStandard Theory (four complex dimensions), four complex quaternion dimension QUeST, and eight complex octonion dimension MOST. This appendix describes the boosts and species in each theory.

In complex quaternion (and octonion) space we must restrict the "time" quaternion (octonion) to be "real-valued" for the purpose of determining the fermion and vector boson spectrums. In this appendix we summarize results of earlier books.

We find that each of the three theories has a "charged" lepton species, a neutral lepton species, an up-type quark species, and a down-type quark species. The octonion quark species, because they are in an 8-dimensional space, have a distinctive higher dimensional feature with a six-dimensional "internal" momentum.

A.1 Four Fermion Species – Unified SuperStandard Theory

In this section[36] we show the free form of each of the four fermion species (charged lepton, neutral lepton, up-type quark, and down-type quark) can be generated from a spin ½ particle spinor at rest using Complex Lorentz group boosts. *A crucial part of this study is the requirement that the relevant Lorentz boosts transform from*

---

[36] See Blaha (2018e) and earlier books by the author for more complete details.

fermion rest frames to "moving" frames with a real-valued energy and a real-valued time. The four species fermion spectrum form is a direct consequence as this section shows.

A.1.1 Matrix Representation of Complex Lorentz Group L_C Boosts

We begin with Complex Lorentz Group (L_C) boosts because they will be crucial in the determination of the equations of motion of various types of spin ½ particles. Defining

$$\omega \mathbf{w} = \mathbf{u_r}\omega_r + i\mathbf{u_i}\omega_i \tag{A.1}$$

we see an L_C boost can be expressed in the form

$$\Lambda_C(\mathbf{v_c}) = \exp[i\omega\hat{\mathbf{w}}\cdot\mathbf{K}] \tag{A.2}$$

where

$$\omega = (\omega_r^2 - \omega_i^2 + 2i\omega_r\omega_i\,\hat{\mathbf{u}}_r\cdot\hat{\mathbf{u}}_i)^{\frac{1}{2}} \tag{A.3}$$
$$\hat{\mathbf{w}} = (\omega_r\hat{\mathbf{u}}_r + i\omega_i\hat{\mathbf{u}}_i)/\omega \tag{A.4}$$

Since $\hat{\mathbf{u}}_r\cdot\hat{\mathbf{u}}_r = 1 = \hat{\mathbf{u}}_i\cdot\hat{\mathbf{u}}_i$

$$\hat{\mathbf{w}}\cdot\hat{\mathbf{w}} = 1 \tag{A.5}$$

The (generally) complex relative velocity is

$$\mathbf{v_c} = \hat{\mathbf{w}}\tanh(\omega) \tag{A.6}$$

We now analytically continue to complex ω and complex unit vectors $\hat{\mathbf{w}}$. The resulting complex generalization will be the matrix form of proper L_C boosts:

$$\Lambda_C(\mathbf{v_c}) = \exp[i\omega\hat{\mathbf{w}}\cdot\mathbf{K}] \equiv \Lambda_C(\omega, \hat{\mathbf{w}})$$

$$= \begin{bmatrix} \cosh(\omega) & -\sinh(\omega)\hat{w}_x & -\sinh(\omega)\hat{w}_y & -\sinh(\omega)\hat{w}_z \\ -\sinh(\omega)\hat{w}_x & 1+(\cosh(\omega)-1)\hat{w}_x^2 & (\cosh(\omega)-1)\hat{w}_x\hat{w}_y & (\cosh(\omega)-1)\hat{w}_x\hat{w}_z \\ -\sinh(\omega)\hat{w}_y & (\cosh(\omega)-1)\hat{w}_x\hat{w}_y & 1+(\cosh(\omega)-1)\hat{w}_y^2 & (\cosh(\omega)-1)\hat{w}_y\hat{w} \\ -\sinh(\omega)\hat{w}_z & (\cosh(\omega)-1)\hat{w}_x\hat{w}_z & (\cosh(\omega)-1)\hat{w}_y\hat{w}_z & 1+(\cosh(\omega)-1)\hat{w}_z^2 \end{bmatrix}$$

(A.7)

Since analytic continuations are unique, the above form for $\Lambda_C(\mathbf{v_c})$ is well-defined and unique. It spans the complete set of proper L_C boosts.

A.1.2 Left-handed Part of LC

If we let

$$\hat{\mathbf{u}}_i = \hat{\mathbf{u}}_r \equiv \hat{\mathbf{u}} \qquad (A.8)$$

so that the vector $\hat{\mathbf{u}}_i$ is parallel to $\hat{\mathbf{u}}_r$, and

$$\omega_i = \pi/2 \qquad (A.9)$$

then $\Lambda_C(\mathbf{v_c})$ becomes a Left-handed L_C boost:

$$\Lambda_C(\mathbf{v_c}) = \Lambda_L(\omega_r, \mathbf{u}) \qquad (A.10)$$

A.1.3 Right-handed part of L_C

If we let

$$\hat{\mathbf{u}}_i = -\hat{\mathbf{u}}_r \equiv -\hat{\mathbf{u}} \qquad (A.11)$$

so that the vector $\hat{\mathbf{u}}_i$ is anti-parallel to $\hat{\mathbf{u}}_r$, and

$$\omega_i = -\pi/2 \qquad (A.12)$$

then $\Lambda_C(\mathbf{v_c})$ becomes a Right-handed L_C boost:

$$\Lambda_C(\mathbf{v_c}) = \Lambda_R(\omega_r, \mathbf{u}) \tag{A.13}$$

as described in Blaha (2007b).

A.1.4 Free Spin ½ Particles – Leptons & Quarks

In this section we begin by developing dynamical equations for spin ½ particles based on the L_C parts. These spin ½ particles are conventional Dirac particles (Majorana particles are also allowed but not discussed), spin ½ tachyons, and "color" versions of both types totaling four species. We will identify leptons and quarks with these fields.

A.1.4.1 First Step - Deriving the Conventional Dirac Equation

In this section we will review a method of obtaining the equation of motion of a particle using a free Dirac equation that is obtained by a Lorentz boost of a spinor wave function of a particle at rest.

In the case of a Lorentz transformation the 4×4 matrix form of a Lorentz transformation of Dirac matrices is

$$S^{-1}(\Lambda(v))\gamma^\nu S(\Lambda(v)) = \Lambda^\nu_\mu(v)\gamma^\mu \tag{A.14}$$

where $S(\Lambda(v))$ is

$$S(\Lambda(v)) = \exp(-i\omega\sigma_{0i}v_i/(2|\mathbf{v}|)) = \exp(-\omega\gamma^0\boldsymbol{\gamma}\cdot\mathbf{v}/(2|\mathbf{v}|))$$
$$= \cosh(\omega/2)I + \sinh(\omega/2)\gamma^0\boldsymbol{\gamma}\cdot\mathbf{p}/|\mathbf{p}| \tag{A.15}$$

with $\omega = \mathrm{arctanh}(|\mathbf{v}|)$, $\cosh(\omega/2) = [(E+m)/(2m)]^{\frac{1}{2}}$ and $\sinh(\omega/2) = |\mathbf{p}|[2m(E+m)]^{-\frac{1}{2}}$. Also

$$S^{-1}(\Lambda(v)) = \gamma^0 S^\dagger(\Lambda(v))\gamma^0 = \exp(\omega\gamma^0\boldsymbol{\gamma}\cdot\mathbf{v}/(2|\mathbf{v}|))$$
$$= \cosh(\omega/2)I - \sinh(\omega/2)\gamma^0\boldsymbol{\gamma}\cdot\mathbf{p}/|\mathbf{p}| \tag{A.16}$$

We begin by defining a generic positive energy plane wave solution of the Dirac equation for a normal fermion particle at rest with rest mass m, *which we take to be the qube bare mass in the absence of interactions*, as

$$\psi(x) = e^{-imt}w(0) \tag{A.17}$$

with w(0) a four component logic spinor column vector. *For a free particle at rest, the rest energy m = m_0, the qube mass.* The wave function satisfies the momentum space Dirac equation for a fermion at rest:

$$(m\gamma^0 - m)e^{-imt}w(0) = 0 \tag{A.18}$$

Subsequently we will use a similar procedure to construct the free tachyonic Dirac equation.

If we now apply S(Λ(v)) we find

$$0 = S(\Lambda(v))(m\gamma^0 - m)e^{-imt}w(0) = [mS(\Lambda(v))\gamma^0 S^{-1}(\Lambda(v)) - m]S(\Lambda(v))w(0)$$

A straightforward evaluation shows

$$mS(\Lambda(v))\gamma^0 S^{-1}(\Lambda(v)) = g_{\mu\nu}p^\mu\gamma^\nu = \not{p} \tag{A.19}$$

where $p^0 = (p^2 + m^2)^{1/2}$, $\mathbf{p} = \gamma m v$, and p = |$\mathbf{p}$|. In addition

$$S(\Lambda(v))w(0) = w(p) \tag{A.20}$$

is a positive energy Dirac spinor. Therefore the Dirac equation for a fermion in motion in momentum space has the form:

$$(\not{p} - m)e^{-ip\cdot x}w(p) = 0 \tag{A.21}$$

where the exponential factor, mt, is also boosted to p·x. Eq. A.21 implies the well-known free, coordinate space Dirac equation:

$$(i\gamma^\mu \partial/\partial x^\mu - m)\psi(x) = 0 \tag{A.22}$$

A.1.4.2 Derivation of the Tachyon Dirac Equation

The Left-handed boost has the form:

$$\Lambda_L(\omega, \mathbf{u}) = \Lambda(\omega + i\pi/2, \mathbf{u}) = \exp[i\omega_L \hat{\mathbf{u}} \cdot \mathbf{K}] \tag{A.23}$$

where $\omega_L = \omega + i\pi/2$ and

$$\cosh(\omega_L) = i \sinh(\omega) = -\gamma = i\gamma_s \tag{A.24}$$
$$\sinh(\omega_L) = i \cosh(\omega) = -\beta\gamma = i\beta\gamma_s$$

with, $\beta = v > 1$, $\gamma_s = (\beta^2 - 1)^{-\frac{1}{2}}$, and $\omega \geq 0$. Thus

$$\sinh(\omega) = \gamma_s \tag{A.25}$$
$$\cosh(\omega) = \beta\gamma_s$$

The corresponding spinor transformation is:

$$S_L(\Lambda_L(\omega, \mathbf{u})) = \exp(-i\omega_L \sigma_{0i} v_i/(2|\mathbf{v}|)) = \exp(-\omega_L \gamma^0 \boldsymbol{\gamma} \cdot \mathbf{v}/(2|\mathbf{v}|))$$
$$= \cosh(\omega_L/2)I + \sinh(\omega_L/2)\gamma^0 \boldsymbol{\gamma} \cdot \mathbf{p}/|\mathbf{p}| \tag{A.26}$$

The inverse transformation is

$$S_L^{-1}(\Lambda_L(\omega, \mathbf{u})) = \gamma^2 \gamma^0 K^{-1} S_L^\dagger K \gamma^0 \gamma^2 = \gamma^2 \gamma^0 S_L^{T} \gamma^0 \gamma^2 = \exp(\omega_L \gamma^0 \boldsymbol{\gamma} \cdot \mathbf{v}/(2|\mathbf{v}|))$$
$$= \cosh(\omega_L/2)I - \sinh(\omega_L/2)\gamma^0 \boldsymbol{\gamma} \cdot \mathbf{p}/|\mathbf{p}| \tag{A.27}$$

where the superscript T denotes the transpose and K is the complex conjugation operator (that also appears in the time-reversal operator). Note that S_L is not unitary just as the equivalent spinor Lorentz transformation $S(\Lambda(v))$ is not unitary.

We can now apply a left-handed superluminal transformation to the generic positive energy plane wave solution of the Dirac equation for a particle of mass m at rest. The result is

$$0 = S_L(\Lambda_L(\omega, \mathbf{u}))(m\gamma^0 - m)e^{-imt}w(0)$$
$$= [mS_L\gamma^0 S_L^{-1} - m]e^{-imt}S_L w(0)$$

where $S_L = S_L(\Lambda_L(\omega, \mathbf{u}))$. After some algebra

$$mS_L\gamma^0 S_L^{-1} = m[\cosh(\omega_L)\gamma^0 - \sinh(\omega_L)\boldsymbol{\gamma}\cdot\mathbf{p}/|\mathbf{p}|]$$
$$= i\gamma^0 E - i\boldsymbol{\gamma}\cdot\mathbf{p} = i\not{p} \tag{A.28}$$

using the tachyon energy and momentum expressions

$$\mathbf{p} = mv\gamma_s \qquad\qquad E = m\gamma_s \tag{A.29}$$

Also

$$S_L w(0) = w_T(p) \tag{A.30}$$

is a tachyon spinor.

The momentum space tachyonic Dirac equation is

$$(i\not{p} - m)e^{ip\cdot x}w_T(p) = 0 \tag{A.31}$$

where p·x = Et – **p·x** after performing a corresponding left-handed superluminal coordinate transformation in the exponential factor. Thus a positive energy wave is transformed into a negative energy wave by the superluminal transformation.

If we apply $i\not{p}$ we find the tachyon mass condition is satisfied

$$-E^2 + \mathbf{p}^2 = m^2 \tag{A.32}$$

Transforming back to coordinate space we obtain the *tachyon Dirac equation*:

$$(\gamma^\mu \partial/\partial x^\mu - m)\psi_T(x) = 0 \tag{A.33}$$

The "missing" factor of i in the first term of eq. A.33 requires the lagrangian to be different from the conventional Dirac lagrangian in order for the lagrangian to be real. The simplest, physically acceptable, free spin ½ tachyon lagrangian density is:

$$\mathcal{L}_T = \psi_T{}^S(\gamma^\mu \partial/\partial x^\mu - m)\psi_T(x) \tag{A.34}$$

where

$$\psi_T{}^S = \psi_T{}^\dagger \, i\gamma^0\gamma^5 \tag{A.35}$$

The corresponding action is

$$I = \int d^4x \mathcal{L}_T \qquad (A.36)$$

A.1.4.3 Complex Space, and 3-Momentum & Real-Valued Energy Fermions (Quarks)

Spinor boost transformations were used in previous sections to develop the dynamical equations for Dirac fields and tachyon fields. In this section we will use L_C spinor boosts to generate additional fermion field dynamical equations.

The form of the L_C spinor boost transformation corresponding to the coordinate transformation is:

$$S_C(\omega, \mathbf{v_c}) = \exp(-i\omega\sigma_{0k}\hat{w}_k/2) = \exp(-\omega\gamma^0\boldsymbol{\gamma}\cdot\hat{\mathbf{w}}/2)$$
$$= \cosh(\omega/2)I + \sinh(\omega/2)\gamma^0\boldsymbol{\gamma}\cdot\hat{\mathbf{w}} \qquad (A.37)$$

The inverse transformation is

$$S_C^{-1}(\omega, \mathbf{v_c}) = \gamma^2\gamma^0 K^{-1}S_C^\dagger K\gamma^0\gamma^2 = \gamma^2\gamma^0 S_C^{\ T}\gamma^0\gamma^2 = \exp(\omega\gamma^0\boldsymbol{\gamma}\cdot\hat{\mathbf{w}}/2)$$
$$= \cosh(\omega/2)I - \sinh(\omega/2)\gamma^0\boldsymbol{\gamma}\cdot\hat{\mathbf{w}} \qquad (A.38)$$

where the superscript T denotes the transpose and K is the complex conjugation operator (that also appears in the time-reversal operator). Note that S_C is not unitary just as in previous cases considered in this appendix.

We now redo the development of spin ½ dynamical equations of motion of earlier sections for this more general case of complex ω and $\hat{\mathbf{w}}$. Again we apply a boost to a Dirac equation for a positive energy plane wave particle of mass m at rest:

$$0 = S_C(\omega, \mathbf{v_c}))(m\gamma^0 - m)e^{-imt}w(0)$$
$$= [mS_C\gamma^0 S_C^{-1} - m]e^{-imt}S_C w(0) \qquad (A.39)$$

where $S_C = S_C(\omega, \hat{\mathbf{w}})$. After some algebra

$$mS_C\gamma^0 S_C^{-1} = m[\cosh(\omega)\gamma^0 - \sinh(\omega)\boldsymbol{\gamma}\cdot\hat{\mathbf{w}}] \qquad (A.40)$$

A.1.4.3.1 CASE 1: PARALLEL REAL AND IMAGINARY RELATIVE VECTORS

If the real and imaginary relative vectors parts of $\hat{\mathbf{w}}$, namely $\hat{\mathbf{u}}_r$ and $\hat{\mathbf{u}}_i$, are parallel, then $\hat{\mathbf{u}}_r\cdot\hat{\mathbf{u}}_i = 1$ and

$$\omega = \omega_r + i\omega_i \qquad (A.41)$$

Eq. A.40 can be re-expressed as

$$mS_C\gamma^0S_C^{-1} = m[\cosh(\omega_r)\cos(\omega_i) + i\sinh(\omega_r)\sin(\omega_i)]\gamma^0 - m[\sinh(\omega_r)\cos(\omega_i) + \\ + i\cosh(\omega_r)\sin(\omega_i)]\gamma\cdot\hat{\mathbf{u}}_r \qquad (A.42)$$

or equivalently

$$mS_C\gamma^0S_C^{-1} = \cos(\omega_i)\gamma\cdot p_r + i\sin(\omega_i)\gamma\cdot p_i \qquad (A.43)$$

where

$$p_r{}^0 = m\cosh(\omega_r) \qquad\qquad p_i{}^0 = m\sinh(\omega_r) \qquad (A.44)$$

and

$$\mathbf{p}_r = m\hat{\mathbf{u}}_r\sinh(\omega_r) \qquad\qquad \mathbf{p}_i = m\hat{\mathbf{u}}_r\cosh(\omega_r) \qquad (A.45)$$

If $\omega_i = 0$, then we recover the momentum space Dirac equation. If $\omega_i = \pi/2$, then we obtain the left-handed momentum space tachyon equation. Since the range of ω_i is [0, ∞> (due to the cut along the real ω-plane axis) eq. A.43 corresponds to the results of the Left-Handed Lorentz boost part discussed earlier.

A.1.4.3.2 CASE 2: ANTI-PARALLEL REAL AND IMAGINARY RELATIVE VECTORS

If the real and imaginary relative vectors parts of $\hat{\mathbf{w}}$, $\hat{\mathbf{u}}_r$ and $\hat{\mathbf{u}}_i$, are anti-parallel $\hat{\mathbf{u}}_r = -\hat{\mathbf{u}}_i$, then $\hat{\mathbf{u}}_r\cdot\hat{\mathbf{u}}_i = -1$ and

$$\omega = \omega_r - i\omega_i \qquad (A.46)$$

We can then express eq. A.40 as

$$mS_C\gamma^0S_C^{-1} = m[\cosh(\omega_r)\cos(\omega_i) - i\sinh(\omega_r)\sin(\omega_i)]\gamma^0 - m[\sinh(\omega_r)\cos(\omega_i) - \\ - i\cosh(\omega_r)\sin(\omega_i)]\gamma\cdot\hat{\mathbf{u}}_r \qquad (A.47)$$

or

$$mS_C\gamma^0S_C^{-1} = \cos(\omega_i)\gamma\cdot p_r - i\sin(\omega_i)\gamma\cdot p_i \qquad (A.48)$$

where

$$p_r{}^0 = m\cosh(\omega_r) \qquad\qquad p_i{}^0 = m\sinh(\omega_r) \qquad (A.49)$$

and

$$\mathbf{p}_r = m\hat{\mathbf{u}}_r\sinh(\omega_r) \qquad\qquad \mathbf{p}_i = m\hat{\mathbf{u}}_r\cosh(\omega_r) \qquad (A.50)$$

If $\omega_i = 0$, then we again recover the momentum space Dirac equation, If $\omega_i = \pi/2$, then we obtain the right-handed momentum space tachyon equation. (The range of ω_i is again $[0, \infty>$.)

Note: Since the matrix elements in the boost depend on $\gamma = (1 - \beta^2)^{-\frac{1}{2}}$ with a singularities at $\beta = \pm 1$, which in turn corresponds to $\omega = \pm\infty$, there is a branch cut along the ω axis in the complex ω-plane. Therefore we point out again the product of three Left-handed transformations is not equivalent to a Right-handed transformation.

A.1.4.3.3 CASE 3: COMPLEXONS: A NEW TYPE OF PARTICLE WITH PERPENDICULAR REAL AND IMAGINARY 3-MOMENTA

   If the real and imaginary relative vectors parts of $\hat{\mathbf{w}}$, namely $\hat{\mathbf{u}}_r$ and $\hat{\mathbf{u}}_i$, are perpendicular, $\hat{\mathbf{u}}_r \cdot \hat{\mathbf{u}}_i = 0$, then

$$\omega = (\omega_r^2 - \omega_i^2)^{\frac{1}{2}} \tag{A.51}$$

Thus ω is either pure real ($\omega_r \geq \omega_i$) or pure imaginary ($\omega_r < \omega_i$).

   The momentum space equation generated by the corresponding L_C spinor boost is

$$\{m\cosh(\omega)\gamma^0 - m\sinh(\omega)\gamma\cdot(\omega_r\hat{\mathbf{u}}_r + i\omega_i\hat{\mathbf{u}}_i)/\omega - m\}e^{-ip\cdot x}w_c(p) = 0 \tag{A.52}$$

Defining the momentum 4-vector

$$p = (p^0, \mathbf{p}) \tag{A.53}$$

where

$$p^0 = m\cosh(\omega) \qquad\qquad \mathbf{p} = \mathbf{p}_r + i\mathbf{p}_i \tag{A.54}$$
$$\mathbf{p}_r = m\omega_r\hat{\mathbf{u}}_r \sinh(\omega)/\omega \quad \mathbf{p}_i = m\omega_i\hat{\mathbf{u}}_i \sinh(\omega)/\omega \tag{A.55}$$

and

$$\mathbf{p}_r \cdot \mathbf{p}_i = 0 \tag{A.56}$$

then we obtain a positive energy Dirac-like equation with complex 3-momentum

$$[p\cdot\gamma - m]e^{-ip\cdot x}w_c(p) = 0$$

or, explicitly, (A.57)

$$[p^0\gamma^0 - (\mathbf{p_r} + i\mathbf{p_i})\cdot\gamma - m]e^{-ip\cdot x}w_c(p) = 0$$

with a complex 3-momentum **p** and the 4-momentum mass shell condition:

$$p^2 = p^{0\,2} - \mathbf{p_r}\cdot\mathbf{p_r} + \mathbf{p_i}\cdot\mathbf{p_i} = m^2 \qquad (A.58)$$

Note

$$|\mathbf{v}| = |\mathbf{p}|/p^0 = [(\mathbf{p_r} + i\mathbf{p_i})\cdot(\mathbf{p_r} + i\mathbf{p_i})]^{\frac{1}{2}}/p^0 = \tanh(\omega) \qquad (A.59)$$

and thus the Lorentz factor

$$\gamma = \cosh(\omega) \qquad (A.60)$$

Eq. A.57 is the momentum space equivalent of the wave equation

$$[i\gamma^0\partial/\partial t + i\gamma\cdot(\nabla_r + i\nabla_i) - m]\psi_C(t, \mathbf{x_r}, \mathbf{x_i}) = 0 \qquad (A.61)$$

where

$$x_c = (t, \mathbf{x_r} - i\mathbf{x_i}) \qquad (A.62)$$

and where the grad operators ∇_r and ∇_i are with respect to $\mathbf{x_r}$ and $\mathbf{x_i}$ respectively. Since $\hat{\mathbf{u}}_r\cdot\hat{\mathbf{u}}_i = 0$, we see that there is a subsidiary condition on the wave function

$$\nabla_r\cdot\nabla_i\,\psi_C(t, \mathbf{x_r}, \mathbf{x_i}) = 0 \qquad (A.63)$$

We will call the particles satisfying eqs. A.62 and A.63 *complexons*. In addition we have the anti-commutation relation

$$\{\gamma\cdot\mathbf{p_r}, \gamma\cdot\mathbf{p_i}\} = 0 \qquad (A.64)$$

which in turn implies

$$\gamma\cdot\nabla_r\gamma\cdot\nabla_i\psi_C(t, \mathbf{x_r}, \mathbf{x_i}) = \gamma\cdot\nabla_i\gamma\cdot\nabla_r\psi_C(t, \mathbf{x_r}, \mathbf{x_i}) = 0 \qquad (A.65)$$

Since $p_r = p_i = 0$ in the particle rest frame prior to the complex group boost, the boosted particle spin 4-vector s^μ satisfies

$$s^\mu p_r{}^\mu = s^\mu p_i{}^\mu = 0 \tag{A.66}$$

Note that s^μ is itself complex[37] and, if the spin points in the z-direction prior to the complex boost, then the boosted s^μ has the form

$$s^\mu = (-\sinh(\omega)\hat{w}_z,\, (0,0,1) + (\cosh(\omega) - 1)\hat{w}_z\hat{\mathbf{w}}) \tag{A.67}$$

with $\hat{\mathbf{w}}$ defined earlier: $\hat{\mathbf{w}} = (\omega_r\hat{\mathbf{u}}_r + i\omega_i\hat{\mathbf{u}}_i)/\omega = \mathbf{p}/(m\sinh(\omega))$.

The momentum 4-vector is defined by

$$p = (p^0, \mathbf{p}) \tag{A.68}$$

where

$$p^0 = m \sinh(\omega) \qquad\qquad \mathbf{p} = \mathbf{p}_r + i\mathbf{p}_i \tag{A.69}$$

with

$$\mathbf{p}_r = m\omega_r\hat{\mathbf{u}}_r \cosh(\omega)/\omega \quad \mathbf{p}_i = m\omega_i\hat{\mathbf{u}}_i \cosh(\omega)/\omega \tag{A.70}$$

and

$$\mathbf{p}_r{\cdot}\mathbf{p}_i = 0 \tag{A.71}$$

then we obtain the complexon tachyon equation

$$[ip{\cdot}\gamma - m]e^{+ip{\cdot}x}w_{cL}(p) = 0 \tag{A.72}$$

with a complex 3-momentum $\mathbf{p}$ and the tachyon 4-momentum mass shell condition:[38]

---

[37] This feature of partons, which is not present in ordinary Dirac particles, might be the source of the discrepancies between theory and experiment in deep inelastic parton spin physics which is based on conventional real parton spins.

[38] Note that the presence of the $\mathbf{p}_i{}^2$ term does not change the tachyon requirement that $\mathbf{p}_r{}^2 \geq m^2$ as seen in the previous cases.

$$p^2 = p^{0\,2} - \mathbf{p_r}^2 + \mathbf{p_i}^2 = -m^2 \tag{A.73}$$

Eq. A.72 is the momentum space equivalent of the wave equation

$$[\gamma^0 \partial/\partial t + \gamma \cdot (\nabla_r + i\nabla_i) - m]\psi_{CL}(t, \mathbf{x_r}, \mathbf{x_i}) = 0 \tag{A.74}$$

or

$$[\gamma \cdot \nabla - m]\psi_{CL}(t, \mathbf{x_r}, \mathbf{x_i}) = 0 \tag{A.75}$$

with the subsidiary condition on the wave function

$$\nabla_r \cdot \nabla_i \; \psi_{CL}(t, \mathbf{x_r}, \mathbf{x_i}) = 0 \tag{A.76}$$

also holds. We note that eq. A.74 is covariant under the real Lorentz group.

A.1.5 Up-type Color Quarks

Up-type quarks are assumed[39] to be fermions with complex 3-momenta - complexons, and an internal color SU(3) symmetry, that satisfy $p^2 = m^2$. Their field equation with a color SU(3) index, denoted a, inserted is

$$[i\gamma^0 \partial/\partial t + i\gamma \cdot (\nabla_r + i\nabla_i) - m]\psi_C{}^a(t, \mathbf{x_r}, \mathbf{x_i}) = 0 \tag{A.77}$$

with the subsidiary condition

$$\nabla_r \cdot \nabla_i \; \psi_C{}^a(t, \mathbf{x_r}, \mathbf{x_i}) = 0 \tag{A.78}$$

The free field solution is:

---

[39] The complexon theory that we develop and use for quark dynamics in the Standard Model is <u>not</u> required. Our SuperStandard Model could use Dirac fermion dynamics for the up-type quarks and tachyon dynamics for down-type quarks. Then the (broken) Left-handed complex Lorentz boosts would have the basic space-time group rather than L_C. We choose to use complexon dynamics for quarks because they have an internal SU(3)-like structure suggestive of color SU(3). More importantly, their spin dynamics is different and thus may resolve the differences between theory and experiment for the deep inelastic parton spin-dependent structure functions.

$$\psi_C^{\ a}(x) = \sum_{\pm s} \int d^3p_r d^3p_i \ N_C(p)\delta(\mathbf{p_r \cdot p_i}/m^2)[b_C(p,a,s)u_C^{\ a}(p,\ s)e^{-i(p\cdot x + p^*\cdot x^*)/2} +$$
$$+ d_C^{\ \dagger}(p,a,s)v_C^{\ a}(p,\ s)e^{+i(p\cdot x + p^*\cdot x^*)/2}] \qquad (A.79)$$

The free Feynman propagator arranged into the form of a spectral integral is

$$iS_C^{\ ab}(x,y) = -\delta^{ab}\int dM \ (i\gamma^0 \partial/\partial x^0 - i(\nabla_r - i\nabla_i)\cdot\gamma + m)\delta'(\nabla_r\cdot\nabla_i/m^2)J(\mathbf{x_i} - \mathbf{y_i}, M^2)\triangle_F(x - y, M)$$
$$(A.80)$$

where

$$\triangle_F(x - y, M) = (2\pi)^{-4}\int d^4p_r \ \exp[-ip^0(x^0 - y^0) + i\mathbf{p_r\cdot(x_r - y_r)}]/(p_r^2 - M^2 + i\varepsilon)$$
$$(A.81)$$

and

$$J(\mathbf{x_i}, M^2) = (2\pi)^{-3}\int d^3p_i \ \delta(M^2 + \mathbf{p_i}^2 - m^2) \ \exp[-i\mathbf{p_i\cdot(x_i - y_i)}] \qquad (A.82)$$
$$= (2\pi)^{-2}|\mathbf{x_i - y_i}|^{-1}\theta(m^2 - M^2)\sin((m^2 - M^2)^{\frac{1}{2}}|\mathbf{x_i - y_i}|)$$

A.1.6 Down-type Color Quarks

Tachyonic complexons with complex 3-momenta, and an internal global SU(3) symmetry, that have mass shell condition $p^2 = -m^A$. Their field equation with a color SU(3) index, denoted a, inserted is

$$[\gamma^0 \partial/\partial t + \gamma\cdot(\nabla_r + i\nabla_i) - m]\psi_{CL}^{\ a}(t, \mathbf{x_r}, \mathbf{x_i}) = 0 \qquad (A.83)$$

with the subsidiary condition on the wave function

$$\nabla_r\cdot\nabla_i \ \psi_{CL}^{\ a}(t, \mathbf{x_r}, \mathbf{x_i}) = 0 \qquad (A.84)$$

Its free field left-handed solution is:

$$\psi_{CLL}^{\ +a}(\mathbf{x_r}, \mathbf{x_i}) = \sum_{\pm s}\int d^2p_r dp^{\dagger}d^3p_i \ N_{CLL}^{\ +}(p)\theta(p^{\dagger})\delta((p_i^3(p^{\dagger} - p^{-})/\sqrt{2} + \mathbf{p_{r\perp}\cdot p_{i\perp}})/m^2)\cdot \qquad (A.85)$$
$$\cdot[b_{CLL}^{\ +}(p,a,s)u_{CLL}^{\ a}(p,a,s)e^{-i(p\cdot x + p^*\cdot x^*)/2} + d_{CLL}^{\ +\dagger}(p,a,s)v_{CLL}^{\ +a}(p,a,s)e^{+i(p\cdot x + p^*\cdot x^*)/2}]$$

and its right-handed solution is

$$\psi_{CLR}^{+a}(x_r, x_i) = \sum_{\pm s} \int d^2p_r dp^+ d^3p_i \, N_{CLR}^{+}(p)\theta(p^+)\delta((p_i^3(p^+ - p^-)/\sqrt{2} + \mathbf{p}_{r\perp}\cdot\mathbf{p}_{i\perp})/m^2)\cdot$$

$$\cdot[b_{CLR}^{+}(p,a,s)u_{CLR}^{+a}(p,a,s)e^{-i(p\cdot x + p^*\cdot x^*)/2} + d_{CLR}^{++}(p,a,s)v_{CLR}^{+a}(p,a,s)e^{+i(p\cdot x + p^*\cdot x^*)/2}]$$

$$(A.86)$$

The free left-handed Feynman propagator arranged into the form of a spectral integral is

$$iS_{CLLF}^{+ab}(x,y) = -\delta^{ab}\int dM \, C^-R^+(\gamma^0\partial/\partial x^0 + (\nabla_r - i\nabla_i)\cdot\gamma - m)R^-C^+\delta'(\nabla_r\cdot\nabla_i/m^2)\cdot$$

$$\cdot J_2(\mathbf{x}_i - \mathbf{y}_i, M^2)\Delta_{FT}(x - y, M) \qquad (A.87)$$

with ∇_r and ∇_i derivatives with respect to $\mathbf{x}_r$ and $\mathbf{x}_i$ and where

$$\Delta_{FT}(x - y, M) = (2\pi)^{-4}\int d^4p_r \exp[-ip^0(x^0 - y^0) + i\mathbf{p}_r\cdot(\mathbf{x}_r - \mathbf{y}_r)]/(p_r^2 + M^2 + i\varepsilon) \qquad (A.88)$$

and

$$J_2(\mathbf{x}_i, M^2) = (2\pi)^{-3}\int d^3p_i \, \delta(M^2 - \mathbf{p}_i^2 - m^2) \exp[-i\mathbf{p}_i\cdot(\mathbf{x}_i - \mathbf{y}_i)] \qquad (A.89)$$

$$= (2\pi)^{-2}|\mathbf{x}_i - \mathbf{y}_i|^{-1}\theta(M^2 - m^2)\sin((M^2 - m^2)^{\frac{1}{2}}|\mathbf{x}_i - \mathbf{y}_i|)$$

A.1.7 Summary: 4 Species of Particles: Leptons and Quarks

A.1.7.1 Charged lepton fermions
    The conventional Dirac equation and solutions.

A.1.7.2 Neutral leptons - Neutrinos
    Simple tachyons with real energy and 3-momentum. Their free field equation is:

$$(\gamma^\mu\partial/\partial x^\mu - m)\psi_T(x) = 0 \qquad (A.90)$$

and their left-handed $\psi_{TL}{}^{+}$ Feynman propagator is:

$$iS^{+}{}_{TLF}(x, y) = \tfrac{1}{2}C^{-}R^{+}\gamma^{0}\!\int\! d^{4}p(2\pi)^{-4}\, p^{+}e^{-ip\cdot(x-y)}/(p^{2} + m^{2} + i\varepsilon) \qquad (A.91)$$

Similarly the light-front Feynman propagator for the right-handed $\psi_{TR}{}^{+}$ tachyon field is

$$iS^{+}{}_{TRF}(x,y) = -\tfrac{1}{2}C^{+}R^{+}\gamma^{0}\!\int\! d^{4}p(2\pi)^{-4}\, p^{+}e^{-ip\cdot(x-y)}/(p^{2} + m^{2} + i\varepsilon) \qquad (A.92)$$

A.1.7.3 Up-type Color Quarks

Up-type quarks are assumed[40] to be fermions with complex 3-momenta - complexons, and an internal color SU(3) symmetry, that satisfy $p^{2} = m^{2}$. Their field equation with a color SU(3) index, denoted a, inserted is

$$[i\gamma^{0}\partial/\partial t + i\gamma\cdot(\nabla_{\mathbf{r}} + i\nabla_{\mathbf{i}}) - m]\psi_{C}{}^{a}(t, \mathbf{x_r}, \mathbf{x_i}) = 0 \qquad (A.93)$$

with the subsidiary condition

$$\nabla_{\mathbf{r}}\cdot\nabla_{\mathbf{i}}\,\psi_{C}{}^{a}(t, \mathbf{x_r}, \mathbf{x_i}) = 0 \qquad (A.94)$$

A.1.7.4 Down-type Color Quarks

Tachyonic complexons with complex 3-momenta, and an internal global SU(3) symmetry, that have mass shell condition $p^{2} = -m^{2}$. Their field equation with a color SU(3) index, denoted a, inserted is

$$[\gamma^{0}\partial/\partial t + \gamma\cdot(\nabla_{\mathbf{r}} + i\nabla_{\mathbf{i}}) - m]\psi_{CL}{}^{a}(t, \mathbf{x_r}, \mathbf{x_i}) = 0 \qquad (A.95)$$

with the subsidiary condition on the wave function

---

[40] The complexon theory that we develop and use for quark dynamics in the Standard Model is not required. Our SuperStandard Model could use Dirac fermion dynamics for the up-type quarks and tachyon dynamics for down-type quarks. Then the (broken) Left-handed complex Lorentz boosts would have the basic space-time group rather than L_C. We choose to use complexon dynamics for quarks because they have an internal SU(3)-like structure suggestive of color SU(3). More importantly, their spin dynamics is different and thus may resolve the differences between theory and experiment for the deep inelastic parton spin-dependent structure functions.

$$\nabla_r \cdot \nabla_i \, \psi_{CL}{}^a(t, \, x_r, \, x_i) = 0 \qquad (A.96)$$

A.2 Four QUeST Fermion Species

QUeST is defined in a 4-dimensional complex quaternion space. We have extracted four complex-valued (eight real-valued) coordinates to form a space-time. These coordinates support a Complex Lorentz group just as the United SuperStandard theory. Therefore the considerations of section A.1 above apply without change.

There are four species of fundamental fermions: "charged" lepton species, neutral lepton species, up-type quark species, and down-type quark species. The nature of each species is the same as in the Unified SuperStandard Theory.

The key relation in complex quaternion space is

$$e^x = e^a \, (\cos{(\|v\|)} + v/\|v\| \, \sin(\|v\|))s \qquad (5.4)$$

in analogy with the similar complex-valued identity used in section A.1.

A.3 MOST Fermion Species

The fermion species in MOST number four despite the complex eight-dimensional nature of the space-time extracted from complex octonion space. The eight complex space-time coordinates support a 7+1 Complex Lorentz group.

There are four types of boosts in 7+1 space-time that boost a particle rest state to a state of motion with a real energy and a real-valued or complex-valued momentum (spatial coordinates):

1. A boost from rest to a frame with real-valued energy and momentum with $p^{02} - p^2 > 0$. A "normal charged lepton-like" fermion.
2. A boost from rest to a frame with real-valued energy and momentum with $p^{02} - p^2 < 0$. A "tachyonic neutral lepton-like" fermion.
3. A boost from rest to a frame with real-valued energy and a complex-valued spatial momentum with $p^{02} - p^2 > 0$. An "up-type quark-like" fermion.
4. A boost from rest to a frame with real-valued energy and momentum with $p^{02} - p^2 < 0$. A "tachyonic down-type quark-like" fermion.

where p^0 is the energy and **p** is a spatial 7-vector. The particle states generated by 1 and 2 are "conventional" 8-dimensional analogues of the 4-dimensional case.

The quark-like cases have an 8-dimensional aspect that prompts us to call them *octoquarks.*in 4-dimensional space-time. In 4-space-time a quark momentum is set by the term

$$\omega w = \mathbf{u_r}\omega_r + i\mathbf{u_i}\omega_i \qquad (A.1)$$

as shown in section A.1. In 8-dimensional space-time the 8-momentum, *hyper-momentum,*[41] is set by

$$\omega w = \mathbf{u_r}\omega_r + i\mathbf{u_i}\omega_i + j\mathbf{u_3}\omega_i + k\mathbf{u_4}\omega_4 + q\mathbf{u_5}\omega_5 + r\mathbf{u_6}\omega_6 \qquad (A.97)$$

where the $\mathbf{u_i}$ are 8-vectors satisfying

$$\mathbf{u_i}\cdot\mathbf{u_j} = \delta_{ij} \qquad (A.98)$$

and i, j, k, q, and r are fundamental octonion units.

Note that a fermion spin s satisfies the 8-vector inner product

$$s\cdot p = 0 \qquad (A.99)$$

or

$$\mathbf{s}\cdot\mathbf{p} = 0 \qquad (A.100)$$

in a manner similar to 4-space-time.

Thus the 7 7-vectors in 8-space-time form an orthonormal set and define quarks with 7-momentum and 7-spin. We call them *octoquarks.*

Normal and tachyonic particles are distinguished in complex 8-dimensional space by the sign of $p^{02} - \mathbf{p}^2$ in a manner similar to 4-space-time. MOST fermions occur in four species just like QUeST fermions and Unified SuperStandard Theory fermions.

There are four species of fundamental fermions in 4-space-time and 8-space-time: "charged" lepton species, neutral lepton species, up-type octoquark species, and

---

[41] As opposed to complex momentum in 4-space-time.

down-type octoquark species. The nature of each species (normal or tachyon, and real-valued or complex spatial momentum) is the same as in the Unified SuperStandard Theory.

A.4 Subgroups of the Complex Lorentz Group

In Blaha (2018e) and earlier books the author found that the restriction to real-valued time in the set of transformations of the Complex Lorentz group led to the presence of the same set of subgroups: U(1), SU(2), and SU(3) as well as other subgroups: U(1), SU(2), and SU(3). We saw the analogous internal symmetry groups existed in the extended Standard Model.

Thus it appears that one can extract the internal group structure of the complex quaternion and the complex octonion spaces by only considering the set of dimensions limited to the real part of the complex time quaternion and the real part of the complex time octonion respectively.

REFERENCES

Bjorken, J. D., Drell, S. D., 1964, *Relativistic Quantum Mechanics* (McGraw-Hill, New York, 1965).

Bjorken, J. D., Drell, S. D., 1965, *Relativistic Quantum Fields* (McGraw-Hill, New York, 1965).

Blaha, S., 1998, *Cosmos and Consciousness* (Pingree-Hill Publishing, Auburn, NH, 1998).

_____, 2002, *A Finite Unified Quantum Field Theory of the Elementary Particle Standard Model and Quantum Gravity Based on New Quantum Dimensions™ & a New Paradigm in the Calculus of Variations* (Pingree-Hill Publishing, Auburn, NH, 2002).

_____, 2003, *A Finite Unified Quantum Field Theory of the Elementary Particle Standard Model and Quantum Gravity Based on New Quantum Dimensions™ and a New Paradigm in the Calculus of Variations* (Pingree-Hill Publishing, Auburn, NH, 2003).

_____, 2004, *Quantum Big Bang Cosmology: Complex Space-time General Relativity, Quantum Coordinates™Dodecahedral Universe, Inflation, and New Spin 0, ½, 1 & 2 Tachyons & Imagyons* (Pingree-Hill Publishing, Auburn, NH, 2004).

_____, 2005a, *Quantum Theory of the Third Kind: A New Type of Divergence-free Quantum Field Theory Supporting a Unified Standard Model of Elementary Particles and Quantum Gravity based on a New Method in the Calculus of Variations* (Pingree-Hill Publishing, Auburn, NH, 2005).

_____, 2005b, *The Metatheory of Physics Theories, and the Theory of Everything as a Quantum Computer Language* (Pingree-Hill Publishing, Auburn, NH, 2005).

_____, 2005c, *The Equivalence of Elementary Particle Theories and Computer Languages: Quantum Computers, Turing Machines, Standard Model, Superstring Theory, and a Proof that Gödel's Theorem Implies Nature Must Be Quantum* (Pingree-Hill Publishing, Auburn, NH, 2005).

_____, 2006a, *The Foundation of the Forces of Nature* (Pingree-Hill Publishing, Auburn, NH, 2006).

_____, 2006b, *A Derivation of ElectroWeak Theory based on an Extension of Special Relativity; Black Hole Tachyons; & Tachyons of Any Spin.* (Pingree-Hill Publishing, Auburn, NH, 2006).

_____, 2007a, *Physics Beyond the Light Barrier: The Source of Parity Violation, Tachyons, and A Derivation of Standard Model Features* (Pingree-Hill Publishing, Auburn, NH, 2007).

_____, 2007b, *The Origin of the Standard Model: The Genesis of Four Quark and Lepton Species, Parity Violation, the ElectroWeak Sector, Color SU(3), Three Visible Generations of Fermions, and One Generation of Dark Matter with Dark Energy* (Pingree-Hill Publishing, Auburn, NH, 2007).

_____, 2008a, *A Direct Derivation of the Form of the Standard Model From GL(16) (Pingree-Hill Publishing, Auburn, NH, 2008).*

_____, 2008b, *A Complete Derivation of the Form of the Standard Model With a New Method to Generate Particle Masses Second Edition* (Pingree-Hill Publishing, Auburn, NH, 2008)

_____, 2009, *The Algebra of Thought & Reality: The Mathematical Basis for Plato's Theory of Ideas, and Reality Extended to Include A Priori Observers and Space-Time Second Edition* (Pingree-Hill Publishing, Auburn, NH, 2009).

_____, 2010a, *Operator Metaphysics: A New Metaphysics Based on a New Operator Logic and a New Quantum Operator Logic that Lead to a Mathematical Basis for Plato's Theory of Ideas and Reality* (Pingree-Hill Publishing, Auburn, NH, 2010).

_____, 2010b, *The Standard Model's Form Derived from Operator Logic, Superluminal Transformations and GL(16)* (Pingree-Hill Publishing, Auburn, NH, 2010).

_____, 2010c, *SuperCivilizations: Civilizations as Superorganisms* (McMann-Fisher Publishing, Auburn, NH, 2010).

_____, 2011a, *21st Century Natural Philosophy Of Ultimate Physical Reality* (McMann-Fisher Publishing, Auburn, NH, 2011).

_____, 2011b, *All the Universe! Faster Than Light Tachyon Quark Starships & Particle Accelerators with the LHC as a Prototype Starship Drive Scientific Edition* (Pingree-Hill Publishing, Auburn, NH, 2011).

_____, 2011c, *From Asynchronous Logic to The Standard Model to Superflight to the Stars* (Blaha Research, Auburn, NH, 2011).

_____, 2012a, *From Asynchronous Logic to The Standard Model to Superflight to the Stars volume 2: Superluminal CP and CPT, U(4) Complex General Relativity and The Standard Model, Complex Vierbein General Relativity, Kinetic Theory, Thermodynamics* (Blaha Research, Auburn, NH, 2012).

_____, 2012b, *Standard Model Symmetries, And Four And Sixteen Dimension Complex Relativity; The Origin Of Higgs Mass Terms* (Blaha Research, Auburn, NH, 2012).

_____, 2013a, *Multi-Stage Space Guns, Micro-Pulse Nuclear Rockets, and Faster-Than-Light Quark-Gluon Ion Drive Starships* (Blaha Research, Auburn, NH, 2013).

_____, 2013b, *The Bridge to Dark Matter; A New Sister Universe; Dark Energy; Inflatons; Quantum Big Bang; Superluminal Physics; An Extended Standard Model Based on Geometry* (Blaha Research, Auburn, NH, 2013).

_____, 2014a, *Universes and Megaverses: From a New Standard Model to a Physical Megaverse; The Big Bang; Our Sister Universe's Wormhole; Origin of the Cosmological Constant, Spatial Asymmetry of the Universe, and its Web of Galaxies; A Baryonic Field* between Universes and Particles; Megaverse Extended Wheeler-DeWitt Equation (Blaha Research, Auburn, NH, 2014).

_____, 2014b, *All the Megaverse! Starships Exploring the Endless Universes of the Cosmos Using the Baryonic Force* (Blaha Research, Auburn, NH, 2014).

_____, 2014c, *All the Megaverse! II Between Megaverse Universes: Quantum Entanglement Explained by the Megaverse Coherent Baryonic Radiation Devices – PHASERs Neutron Star Megaverse Slingshot Dynamics Spiritual and UFO Events, and the Megaverse Microscopic Entry into the Megaverse* (Blaha Research, Auburn, NH, 2014).

_____, 2015a, *PHYSICS IS LOGIC PAINTED ON THE VOID: Origin of Bare Masses and The Standard Model in Logic, U(4) Origin of the Generations, Normal and Dark Baryonic Forces, Dark Matter, Dark Energy, The Big Bang, Complex General Relativity, A Megaverse of Universe Particles* (Blaha Research, Auburn, NH, 2015).

_____, 2015b, *PHYSICS IS LOGIC Part II: The Theory of Everything, The Megaverse Theory of Everything, U(4)⊗U(4) Grand Unified Theory (GUT), Inertial Mass = Gravitational Mass, Unified Extended Standard Model and a New Complex General Relativity with Higgs Particles, Generation Group Higgs Particles* (Blaha Research, Auburn, NH, 2015).

_____, 2015c, *The Origin of Higgs ("God") Particles and the Higgs Mechanism: Physics is Logic III, Beyond Higgs – A Revamped Theory With a Local Arrow of Time, The Theory of Everything Enhanced, Why Inertial Frames are Special, Universes of the Mind* (Blaha Research, Auburn, NH, 2015).

_____, 2015d, *The Origin of the Eight Coupling Constants of The Theory of Everything: U(8) Grand Unified Theory of Everything (GUTE), S^8 Coupling Constant Symmetry, Space-Time Dependent Coupling Constants, Big Bang Vacuum Coupling Constants, Physics is Logic IV* (Blaha Research, Auburn, NH, 2015).

_____, 2016a, *New Types of Dark Matter, Big Bang Equipartition, and A New U(4) Symmetry in the Theory of Everything: Equipartition Principle for Fermions, Matter is 83.33% Dark,*

Penetrating the Veil of the Big Bang, Explicit QFT Quark Confinement and Charmonium, Physics is Logic V (Blaha Research, Auburn, NH, 2016).

_____, 2016b, *The Periodic Table of the 192 Quarks and Leptons in The Theory of Everything: The U(4) Layer Group, Physics is Logic VI* (Blaha Research, Auburn, NH, 2016).

_____, 2016c, *New Boson Quantum Field Theory, Dark Matter Dynamics, Dark Matter Fermion Layer Mixing, Genesis of Higgs Particles, New Layer Higgs Masses, Higgs Coupling Constants, Non-Abelian Higgs Gauge Fields, Physics is Logic VII* (Blaha Research, Auburn, NH, 2016).

_____, 2016d, *Unification of the Strong Interactions and Gravitation: Quark Confinement Linked to Modified Short-Distance Gravity; Physics is Logic VIII* (Blaha Research, Auburn, NH, 2016).

_____, 2016e, *MoND: Unification of the Strong Interactions and Gravitation II, Quark Confinement Linked to Large-Scale Gravity, Physics is Logic IX* (Blaha Research, Auburn, NH, 2016).

_____, 2016f, *CQ Mechanics: A Unification of Quantum & Classical Mechanics, Quantum/Semi-Classical Entanglement, Quantum/Classical Path Integrals, Quantum/Classical Chaos* (Blaha Research, Auburn, NH, 2016).

_____, 2016g, *GEMS: Unified Gravity, ElectroMagnetic and Strong Interactions: Manifest Quark Confinement, A Solution for the Proton Spin Puzzle, Modified Gravity on the Galactic Scale* (Pingree Hill Publishing, Auburn, NH, 2016).

_____, 2016h, *Unification of the Seven Boson Interactions based on the Riemann-Christoffel Curvature Tensor* (Pingree Hill Publishing, Auburn, NH, 2016).

_____, 2017a, *Unification of the Eleven Boson Interactions based on 'Rotations of Interactions'* (Pingree Hill Publishing, Auburn, NH, 2017).

_____, 2017b, *The Origin of Fermions and Bosons, and Their Unification* (Pingree Hill Publishing, Auburn, NII, 2017).

_____, 2017c, *Megaverse: The Universe of Universes* (Pingree Hill Publishing, Auburn, NH, 2017).

_____, 2017d, *SuperSymmetry and the Unified SuperStandard Model* (Pingree Hill Publishing, Auburn, NH, 2017).

_____, 2017e, *From Qubits to the Unified SuperStandard Model with Embedded SuperStrings: A Derivation* (Pingree Hill Publishing, Auburn, NH, 2017).

_____, 2017f, *The Unified SuperStandard Model in Our Universe and the Megaverse: Quarks, ... ,* (Pingree Hill Publishing, Auburn, NH, 2017).

_____, 2018a, *The Unified SuperStandard Model and the Megaverse SECOND EDITION A Deeper Theory based on a New Particle Functional Space that Explicates Quantum Entanglement Spookiness (Volume 1)* (Pingree Hill Publishing, Auburn, NH, 2018).

_____, 2018b, *Cosmos Creation: The Unified SuperStandard Model, Volume 2, SECOND EDITION* (Pingree Hill Publishing, Auburn, NH, 2018).

_____, 2018c, *God Theory (*Pingree Hill Publishing, Auburn, NH, 2018).

_____, 2018d, *Immortal Eye: God Theory: Second Edition* (Pingree Hill Publishing, Auburn, NH, 2018).

_____, 2018e, *Unification of God Theory and Unified SuperStandard Model THIRD EDITION* (Pingree Hill Publishing, Auburn, NH, 2018).

_____, 2019a, *Calculation of: QED α = 1/137, and Other Coupling Constants of the Unified SuperStandard Theory* (Pingree Hill Publishing, Auburn, NH, 2019).

_____, 2019b, *Coupling Constants of the Unified SuperStandard Theory SECOND EDITION* (Pingree Hill Publishing, Auburn, NH, 2019).

_____, 2019c, *New Hybrid Quantum Big_Bang–Megaverse_Driven Universe with a Finite Big Bang and an Increasing Hubble Constant* (Pingree Hill Publishing, Auburn, NH, 2019).

_____, 2019d, *The Universe, The Electron and The Vacuum* (Pingree Hill Publishing, Auburn, NH, 2019).

_____, 2019e, *Quantum Big Bang – Quantum Vacuum Universes (Particles)* (Pingree Hill Publishing, Auburn, NH, 2019).

_____, 2019f, *The Exact QED Calculation of the Fine Structure Constant Implies ALL 4D Universes have the Same Physics/Life Prospects* (Pingree Hill Publishing, Auburn, NH, 2019).

_____, 2019g, *Unified SuperStandard Theory and the SuperUniverse Model: The Foundation of Science* (Pingree Hill Publishing, Auburn, NH, 2019).

_____, 2020, *Quaternion Unified SuperStandard Theory (The QUeST) and Megaverse Octonion SuperStandard Theory (MOST)* (Pingree Hill Publishing, Auburn, NH, 2020).

Eddington, A. S., 1952, *The Mathematical Theory of Relativity* (Cambridge University Press, Cambridge, U.K., 1952).

Fant, Karl M., 2005, *Logically Determined Design: Clockless System Design With NULL Convention Logic* (John Wiley and Sons, Hoboken, NJ, 2005).

Feinberg, G. and Shapiro, R., 1980, *Life Beyond Earth: The Intelligent Earthlings Guide to Life in the Universe* (William Morrow and Company, New York, 1980).

Gelfand, I. M., Fomin, S. V., Silverman, R. A. (tr), 2000, *Calculus of Variations* (Dover Publications, Mineola, NY, 2000).

Giaquinta, M., Modica, G., Souchek, J., 1998, *Cartesian Coordinates in the Calculus of Variations* Volumes I and II (Springer-Verlag, New York, 1998).

Giaquinta, M., Hildebrandt, S., 1996, *Calculus of Variations* Volumes I and II (Springer-Verlag, New York, 1996).

Gradshteyn, I. S. and Ryzhik, I. M., 1965, *Table of Integrals, Series, and Products* (Academic Press, New York, 1965).

Heitler, W., 1954, *The Quantum Theory of Radiation* (Clarendon Press, Oxford, UK, 1954).

Huang, Kerson, 1992, *Quarks, Leptons & Gauge Fields 2nd Edition* (World Scientific Publishing Company, Singapore, 1992).

Jost, J., Li-Jost, X., 1998, *Calculus of Variations* (Cambridge University Press, New York, 1998).

Kaku, Michio, 1993, *Quantum Field Theory*, (Oxford University Press, New York, 1993).

Kirk, G. S. and Raven, J. E., 1962, *The Pre-Socratic Philosophers* (Cambridge University Press, New York, 1962).

Landau, L. D. and Lifshitz, E. M., 1987, *Fluid Mechanics 2nd Edition*, (Pergamum Press, Elmsford, NY, 1987).

Misner, C. W., Thorne, K. S., and Wheeler, J. A., 1973, *Gravitation* (W. H. Freeman, New York, 1973).

Rescher, N., 1967, *The Philosophy of Leibniz* (Prentice-Hall, Englewood Cliffs, NJ, 1967).

Rieffel, Eleanor and Polak, Wolfgang, 2014, *Quantum Computing* (MIT Press, Cambridge, MA, 2014).

Riesz, Frigyes and Sz.-Nagy, Béla, 1990, *Functional Analysis* (Dover Publications, New York, 1990).
Sagan, H., 1993, *Introduction to the Calculus of Variations* (Dover Publications, Mineola, NY, 1993).

Sakurai, J. J., 1964, *Invariance Principles and Elementary Particles* (Princeton University Press, Princeton, NJ, 1964).

Sorokin, Pitirim, 1941, *Social and Cultural Dynamics* (Porter Sargent Publishers, Boston, MA, 1941).

Streater, R. F. and Wightman, A. S., 2000, *PCT, Spin, Statistics, and All That* (Princeton University Press, Princeton, NJ 2000).

Weinberg, S., 1972, *Gravitation and Cosmology* (John Wiley and Sons, New York, 1972).

Weinberg, S., 1995, *The Quantum Theory of Fields Volume I* (Cambridge University Press, New York, 1995).

Weinberg, S., 2000, *The Quantum Theory of Fields Volume III Supersymmetry* (Cambridge University Press, New York, 2000).

Weyl, H., 1950, *Space, Time, Matter* (Dover, New York, 1950).

Weyl, H., (Tr. S. Pollard et al), 1987, *The Continuum* (Dover Publications, New York, 1987).

INDEX

About the Author

Stephen Blaha is a well-known Physicist and Man of Letters with interests in Science, Society and civilization, the Arts, and Technology. He had an Alfred P. Sloan Foundation scholarship in college. He received his Ph.D. in Physics from Rockefeller University. He has served on the faculties of several major universities. He was also a Member of the Technical Staff at Bell Laboratories, a manager at the Boston Globe Newspaper, a Director at Wang Laboratories, and President of Blaha Software Inc. and of Janus Associates Inc. (NH).

Among other achievements he was a co-discoverer of the "r potential" for heavy quark binding developing the first (and still the only demonstrable) non-Abelian gauge theory with an "r" potential; first suggested the existence of topological structures in superfluid He-3; first proposed Yang-Mills theories would appear in condensed matter phenomena with non-scalar order parameters; first developed a grammar-based formalism for quantum computers and applied it to elementary particle theories; first developed a new form of quantum field theory without divergences (thus solving a major 60 year old

problem that enabled a unified theory of the Standard Model and Quantum Gravity without divergences to be developed); first developed a formulation of complex General Relativity based on analytic continuation from real space-time; first developed a generalized non-homogeneous Robertson-Walker metric that enabled a quantum theory of the Big Bang to be developed without singularities at t = 0; first generalized Cauchy's theorem and Gauss' theorem to complex, curved multi-dimensional spaces; received Honorable Mention in the Gravity Research Foundation Essay Competition in 1978; first developed a physically acceptable theory of faster-than-light particles; first derived a composition of extrema method in the Calculus of Variations; first quantitatively suggested that inflationary periods in the history of the universe were not needed; first proved Gödel's Theorem implies Nature must be quantum; provided a new alternative to the Higgs Mechanism, and Higgs particles, to generate masses; first showed how to resolve logical paradoxes including Gödel's Undecidability Theorem by developing Operator Logic and Quantum Operator Logic; first developed a quantitative harmonic oscillator-like model of the life cycle, and interactions, of civilizations; first showed how equations describing superorganisms also apply to civilizations. A recent book shows his theory applies successfully to the past 14 years of history and to *new* archaeological data on Andean and Mayan civilizations as well as Early Anatolian and Egyptian civilizations.

He first developed an axiomatic derivation of the form of The Standard Model from geometry – space-time properties – The Unified SuperStandard Theory. It unifies all the known forces of Nature. It also has a Dark Matter sector that includes a Dark ElectroWeak sector with Dark doublets and Dark gauge interactions. It uses quantum coordinates to remove infinities that crop up in most interacting quantum field theories and additionally to remove the infinities that appear in the Big Bang and generate inflationary growth of the universe. It shows gravity has a MOND-like form without sacrificing Newton's Laws. It relates the interactions of the MOND-like sector of gravity with the r-potential of Quark Confinement. The axioms of the theory lead to the question of their origin. We suggest in the preceding edition of this book it can be attributed to an

entity with God-like properties. We explore these properties in "God Theory" and show they predict that the Cosmos exists forever although individual universes (or incarnations of our universe) "come and go." Several other important results emerge from God Theory such a functionally triune God. The Unified SuperStandard Theory has many other important parts described in the Current Edition of *The Unified SuperStandard Theory* and expanded in subsequent volumes.

Blaha has had a major impact on a succession of elementary particle theories: his Ph.D. thesis (1970), and papers, showed that quantum field theory calculations to all orders in ladder approximations could not give scaling deep inelastic electron-nucleon scattering. He later showed the eigenvalue equation for the fine structure constant α in Johnson-Baker-Willey QED had a zero at $\alpha = 1$ not $1/137$ by solving the Schwinger-Dyson equations to all orders in an approximation that agreed with exact results to 4th order in α thus ending interest in this theory. In 1979 at Prof. Ken Johnson's (MIT) suggestion he calculated the proton-neutron mass difference in the MIT bag model and found the result had the wrong sign reducing interest in the bag model. These results all appear in Physical Review papers. In the 2000's he repeatedly pointed out the shortcomings of SuperString theory and showed that The Standard Model's form could be derived from space-time geometry by an extension of Lorentz transformations to faster than light transformations. This deeper space-time basis greatly increases the possibility that it is part of THE fundamental theory. Recently, Blaha showed that the Weak interactions differed significantly from the Strong, electromagnetic and gravitation interactions in important respects while these interactions had similar features, and suggested that ElectroWeak theory, which is essentially a glued union of the Weak interactions and Electromagnetism, possibly modulo unknown Higgs particle features, be replaced by a unified theory of the other interactions combined with a stand-alone Weak interaction theory. Blaha also showed that, if Charmonium calculations are taken seriously, the Strong interaction coupling constant is only a factor of five larger than the electromagnetic coupling constant, and thus Strong interaction perturbation theory would make sense and yield physically meaningful results.

In graduate school (1965-71) he wrote substantial papers in elementary particles and group theory: The Inelastic E- P Structure Functions in a Gluon Model. Phys. Lett. B40:501-502,1972; Deep-Inelastic E-P Structure Functions In A Ladder Model With Spin 1/2 Nucleons, Phys.Rev. D3:510-523,1971; Continuum Contributions To The Pion Radius, Phys. Rev. 178:2167-2169,1969; Character Analysis of U(N) and SU(N), J. Math. Phys. <u>10,</u> 2156 (1969); and The Calculation of the Irreducible Characters of the Symmetric Group in Terms of the Compound Characters, (Published as Blaha's Lemma in D. E. Knuth's book: *The Art of Computer Programming Vols. 1 – 4*).

In the early 1980's Blaha was also a pioneer in the development of UNIX for financial, scientific and Internet applications: benchmarked UNIX versions showing that block size was critical for UNIX performance, developing financial modeling software, starting database benchmarking comparison studies, developing Internet-like UNIX networking (1982) and developing a hybrid shell programming technique (1982) that was a precursor to the PERL programming language. He was also the manager of the AT&T ten-year future products development database. His work helped lead to commercial UNIX on computers such as Sun Micros, IBM AIX minis, and Apple computers.

In the 1980's he pioneered the development of PC Desktop Publishing on laser printers. and was nominated for three "Awards for Technical Excellence" in 1987 by PC Magazine for PC software products that he designed and developed.

He has developed a theory of Megaverses – actual universes of which our universe is one – with quantum particle-like properties based on the Wheeler-DeWitt equation of Quantum Gravity. He has developed a theory of a baryonic force, which had been conjectured many years ago, and estimated the strength of the force based on discrepancies in measurements of the gravitational constant G. This force, operative in D-dimensional space, can be used to escape from our universe in "uniships" which are the equivalent of the faster-than-light starships proposed in the author's earlier books. Thus travel to other universes, as well as to other stars is possible.

Blaha also considered the complexified Wheeler-DeWitt equation and showed that its limitation to real-valued coordinates and metrics generated a Cosmological Constant in the Einstein equations.

Recently he calculated the QED Fine Structure Constant exactly to the experimentally known 13 places. He also used the same approach to approximately calculate the Weak interaction SU(2) and Strong interaction SU(3) coupling constants successfully. Based on the origin of all coupling constants in quantum field theoretic vacuum polarization effects he suggested all universes would have the same interactions and consequently the same Physics, Chemistry and Biology. Thus all universes would be trivially Anthropic and capable of Life. Going further he suggested universes are particles and that universe expansion is co9mpletely analogous to vacuum polarization of a universe particle due to a gauge vector universe interaction. Universe expansion as a function of time is the fourier transform of universe vacuum polarization. This feature was demonstrated by almost exact (when compared to the universe scale factor) calculation of the small times universe scale factor in perturbation theory.

The author has also recently written a series of books on the serious problems of the United States and their solution as well as a book on the decline of Mankind that will follow from current social and genetic trends in Mankind.

In the past twelve years Dr. Blaha has written over 40 books on a wide range of topics. Some recent major works are: *From Asynchronous Logic to The Standard Model to Superflight to the Stars*, *All the Universe!*, *SuperCivilizations: Civilizations as Superorganisms*, *America's Future: an Islamic Surge*, *ISIS, al Qaeda, World Epidemics, Ukraine, Russia-China Pact, US Leadership Crisis*, *The Rises and Falls of Man – Destiny – 3000 AD: New Support for a Superorganism MACRO-THEORY of CIVILIZATIONS From CURRENT WORLD TRENDS and NEW Peruvian, Pre-Mayan, Mayan, Anatolian, and Early Egyptian Data, with a Projection to 3000 AD*, and *Mankind in Decline: Genetic Disasters, Human-Animal Hybrids, Overpopulation, Pollution, Global Warming, Food and Water Shortages, Desertification, Poverty, Rising Violence, Genocide, Epidemics, Wars, Leadership Failure*.

He has taught approximately 4,000 students in undergraduate, graduate, and postgraduate corporate education courses primarily in major universities, and large companies and government agencies.

www.ingramcontent.com/pod-product-compliance
Lightning Source LLC
Chambersburg PA
CBHW082008190326
41458CB00010B/3121